TRACER METHODS FOR IN VIVO KINETICS

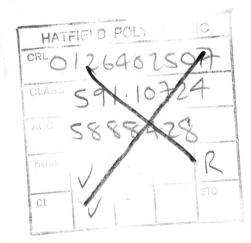

TRACER METHODS FOR IN VIVO KINETICS

THEORY AND APPLICATIONS

Reginald A. Shipley

VETERANS ADMINISTRATION HOSPITAL
CLEVELAND, OHIO
AND
CASE WESTERN RESERVE UNIVERSITY

Richard E. Clark

VETERANS ADMINISTRATION HOSPITAL
CLEVELAND, OHIO

 ACADEMIC PRESS New York and London   1972

ACADEMIC PRESS, INC.
111 Fifth Avenue, New York, New York 10003

United Kingdom Edition published by
ACADEMIC PRESS, INC. (LONDON) LTD.
24/28 Oval Road, London NW1

LIBRARY OF CONGRESS CATALOG CARD NUMBER: 73-187233

PRINTED IN THE UNITED STATES OF AMERICA

CONTENTS

CHAPTER 3. Compartment Analysis: Three-Pool Open Systems

CHAPTER 4. Compartment Analysis: Four or More Pools

CHAPTER 5. Stochastic Analysis: The Stewart–Hamilton Equation

CHAPTER 6. Stochastic Analysis: Rates of Production, Disposal, Secretion, and Conversion; Clearance

CHAPTER 7. Stochastic Analysis: Mean Transit Time, Mass, Volume

CHAPTER 8. Closed Systems, Cumulative Loss, Sinks

CHAPTER 9.   Constant Infusion of Tracer

CHAPTER 10.   Nonsteady State

CHAPTER 11.   Circulation Rate Measured by Tissue Saturation or Desaturation

CHAPTER 12.   Various Approximations from Curves

APPENDIX I.   Derivation of the Formula for Simple Exponential Loss

APPENDIX II.   Mathematical Solution of Multicompartment Models in Steady State

APPENDIX III.   Relation of a Rate Ratio to Vertical Shift in t_{max}

APPENDIX IV.   Deconvolution by Numerical Sequence Approximation

APPENDIX V.   Some Definite Integrals and Derivatives

References

PREFACE

The purpose of this book is to bring together for explanation and evaluation the variety of working formulas called upon in applying tracer methods for kinetic studies *in vivo*. Examination reveals that many formulations, in what at first glance may appear a confusing array, in reality rest upon relatively few related basic concepts, even though the intended use may be for such dissimilar purposes as measurement of cardiac output or estimation of rate of production of a hormone. Readers encountering the wide range of formulas expressed in varying terms and symbols may, understandably, fall victim to a sense of confusion bordering on disenchantment or, at the least, be reduced to uneasy acceptance, on faith, of mathematical development and argument. In order to alleviate this confusion we define the basic concepts, derive the pertinent equations, and evaluate each working formula to ensure its proper application within the set of experimental conditions at hand. Included is an appraisal of inherent potential sources of error encountered in the various applications.

In the interest of the biomedical scientist who as often as not may be uncomfortable when confronted with involved algebra, let alone calculus, we have attempted to maintain as simple and informal an approach as possible in deriving equations. Rigorous and complex qualifying generalities with attending proliferation of symbols have been avoided when the practical purpose—an orderly and straightforward development—is not compromised. We believe that nonmathematicians should welcome such an explanation of concepts supporting various formulations and also appreciate the simple representative numerical examples which are included. Many of the sequences of steps included in the text, whether algebraic or numerical, would ordinarily be omitted from research reports or mathematical treatises; yet we feel certain that many manipulations which are perhaps inelegant and obvious to mathematicians may be helpful to others. In this spirit they are retained here.

Although considerable attention is given stochastic or probabalistic methods as applied to "black box" systems, explicit compartmental models are regularly invoked if for no other reason than to illustrate working formulas in a tangible context and to provide a framework for illustrative arithmetic. We believe that the generous assortment of model systems depicted in the illustrations will assist in the understanding of concepts which may be elusive if sole reliance is placed on abstract mathematical argument.

LIST OF SYMBOLS

A Area under a curve for a given function from t_1 to t_2, or $t = 0$ to $t = \infty$, e.g., $\int_0^\infty f(t)\, dt$.

a, b, c, d, etc.   Designation of specific pools, usually as subscripts. Pool a ordinarily is the labeled pool and the others are secondary pools. Where double labeling requires that other pools also be labeled, this is noted in the text.

α (alpha)   Specific activity of pooled excreted material collected over a period of time sufficiently long to permit recovery of nearly all tracer destined to leave by this route.

C A constant (except electric capacitance (Chapter 4)).

c A general symbol for concentration in water (e.g., blood or plasma) or in tissue. As indicated in the text it may apply to either solute (tracee) or tracer. In case both types of concentration appear in the same formula, that for tracer is designated c', c_a', etc. The symbol c usually appears with a subscript. See the next four entries.

c_a, c_b, etc.   Concentration in pool a, pool b, etc.

c_{a0} Concentration in pool a at zero time.

$c_b{}^a$, $c_a{}^b$, etc.	Concentration of tracer in pool b when dose was to pool a; in pool a when dose was to pool b, etc.
c_A, c_V	Concentration of tracer in arterial or in venous blood, respectively.
Cl	Clearance, expressed as volume nominally cleared of solute per unit time. Cl′ is used for clearance of tracer when that for solute appears in the same formula.
D	Total activity in a single dose of tracer.
D^a, D^b	Dose *to* pool a, dose *to* pool b.
Δt (delta t)	A very small interval of time approaching zero duration.
DR	Irreversible disposal rate of a particular species from a system, e.g., DR_a is disposal rate of species a.
E	Subscript denoting equilibrium.
E	Electric charge, Chapter 4.
e	Natural log base.
F	Rate of transfer (flow) of unlabeled material, e.g., in milliliters per minute or milligrams per minute.
F_{ba}, F_{oa}, F_{ao}, etc.	Rate of transfer *to* pool b from a; to the outside of the system from pool a; to pool a from the outside, etc.
f	Rate of blood flow as milliliters per gram of tissue.
$f(t)$	Any mathematical function of the independent variable, t.
g_1, g_2, etc.	Separate exponential slopes comprising a complex exponential curve.
H_1, H_2, etc.	Coefficients (intercepts) of the separate terms of a complex exponential curve from the labeled pool after observed intercepts (I) are normalized in terms of fraction of their total. Example $H_1 = I_1/(I_1 + I_2 + \cdots I_n)$.
$h(t)$	Transport function: Fraction of dose of tracer lost from a system per unit time as a function of time.
I	An intercept of an extrapolated slope of a complex exponential curve. It also is one of the coefficients in the equation for the curve.
i, j	Subscripts of generality, e.g., A_i means *any given* area; F_{ij} means the rate of movement to any pool, i, from any pool, j.
K_1, K_2, etc.	Coefficients (intercepts) of the separate terms of the complex exponential curve for quantity of tracer in pool b when amount is expressed as fraction of dose to labeled pool a.
k	A rate constant of transfer from a pool in terms of fraction of total content moving per unit time.
k_{ba}, k_{bc}, etc.	See F for definition of subscripts.

k_{aa}, k_{bb}, etc.	Sum of all rate constants of output from pool a, pool b, etc.
L_1, L_2, etc.	Coefficients for curve from pool c (as defined for pool b).
ln	Log to the base e.
$\mathscr{L}\{\ \}$	A Laplace transform.
λ (lambda)	Partition coefficient for a gas: quantity per gram of tissue versus quantity per milliliter of blood.
M_1, M_2, etc.	Coefficients for a curve from pool d (as defined for pool b).
m	Mass (weight) of tissue.
N, or n	Any integer such as the nth member of a series, or the number in a series.
0	Subscript denoting zero time, e.g., t_0 ; q_{a0} is quantity of tracer in pool a at zero time.
o	Outside the system, e.g., F_{ao} is rate of transfer of tracee to pool a from the outside.
P	Quantity of tracer in a priming dose preceding constant infusion.
PR	Production rate of a specific species, e.g., PR_a is production rate of species a (new to the system).
p	A "dummy variable" in a Laplace transform.
q	Quantity of tracer, e.g., counts per minute in a pool or specified space denoted by subscript.
q_a, q_b, etc.	Quantity of tracer in pool a, pool b, etc.
q_{a0}	Quantity of tracer present in pool a at zero time. (Numerically the same as D introduced to pool a.)
$q_b{}^a, q_a{}^b$, etc.	Quantity of tracer in pool b when a single dose was introduced to pool a; quantity of tracer in pool a when dose was to pool b, etc.
Q	Quantity of unlabeled material (tracee) in a pool or space, e.g., Q_a is quantity in pool a.
R	Reading of a radiation detector. (Also electrical resistance, Chapter 4).
r	Rate of movement (or infusion) of tracer as units per unit time.
$S(t)$	Probability function of arrival time of tracer.
SA	Specific activity as units of tracer per unit weight of natural atoms of the same species.
SA_a	SA of pool a, etc.
T	A specific interval of time.
$T_{1/2}$	Half time. Time required for q or SA to decline by half when the curve is of simple exponential type.

T_{mean}	Mean time, e.g., mean time for loss of tracer.
t	Time, as an independent variable; a specific point in time.
(t)	A "function of time," e.g., $q_a(t)$ is amount of radio-activity in pool a (dependent variable) as a function of time (independent variable). (Frequently omitted when a variable is obviously a function of time.)
t_0	Zero time.
t_{max}	Point in time where a curve is at maximum height.
τ (tau)	A time interval on a special subscale in the convolution integral (Chapter 12).
U	A complex denominator of specified constants.
V	Units of liquid volume.
ω (omega)	Subscript (e.g., t_ω, q_ω denoting a value at a point in time terminating an interval during which observations are made).
x, y, z	Variables as defined when used.
∞	At infinite time.

Operator Signs

$\approx$	Approximately equal to.
$\sum$	Summation of.
xy or $(x)(y)$ or $[x][y]$ or $x \cdot y$	x multiplied by y, except that (t) is always "as a function of time" and $(t - \tau)$ is "as a function of $t - \tau$."
!	Factorial. For example 3! is $3 \cdot 2 \cdot 1$, and 4! is $4 \cdot 3 \cdot 2 \cdot 1$.

CHAPTER 1

COMPARTMENT ANALYSIS: A SINGLE POOL, REAL OR BY LUMPING

Compartments or Pools

1. Tracers such as radioactive atoms are assumed to behave chemically and physiologically exactly like their natural counterpart atoms save for slight effects of difference in mass. Because the dose can be very small in terms of the number of existing natural atoms, the added material does not perturb the system under observation. Tracers have several potential uses in the intact animal. One of these is to study pathways of chemical conversion by identifying tracer in product after introduction into a precursor. If such a pathway is already known, tracer may serve to assess *rate* of conversion. The animal body may be viewed as an assortment of pools or compartments each made up of identical molecules which tend, more or less, to be enclosed by anatomic boundaries. For example, glucose resides for the most part in extracellular fluid. Body pools tend to remain constant in size while undergoing replacement by input equal to output. This dynamic equilibrium is known as *steady state*. Such a state will be assumed for all analyses presented in this book unless otherwise noted. (See Chapter 10 for nonsteady state and more explicit definitions.) *Compartment analysis* is based on the assumption that specific pools can be identified and that discharge of tracer therefrom

can be described by exponential equations. Tracer can be delivered to a pool system as a single, abruptly administered dose, or delivered over an extended period as by continuous infusion at a constant rate. For compartment analysis the single dose technique is the most useful. Except in Chapter 9 and where otherwise noted the analytic approach will be that for a single dose.
2.   In addition to measuring rates of chemical transfer or of physical transport such as blood flow or molecular diffusion, compartment analysis also is concerned with assessment of *pool size*, i.e., the mass of natural material (or volume where appropriate) which constitutes the pool. A very important concept is that of rate of *fractional loss* from a pool. The fraction of tracer lost per unit time is known as a fractional rate constant or *rate constant*. If glucose is assumed to constitute a body pool and this pool is labeled with ^{14}C-glucose given as a single dose intravenously, then a curve of specific activity (SA) as units of radioactivity per milligram of glucose carbon is plotted against time, how can this SA curve of declining activity be used to measure the rate constant of glucose loss or the *rate* of loss (and replacement) as milligrams per minute? What is the weight of glucose carbon in the pool? In the sections which follow, these questions will be considered for a *pure* single pool of homogeneous atoms. A pure single pool means that a specifically defined compartment has no side connections to other pools which participate in interchange of tracer with the pool under observation. The truth is that no such pool exists in the animal body, although a compromise sometimes will permit this assumption in a specific instance. In any case, single-pool kinetics must be understood before more complex systems can be examined.

Single-Pool Kinetics

RATE VERSUS RATE CONSTANT

Natural (tracee) atoms and rate constant

3.   A pool of fluid (Figure 1A) will serve as the first illustration. It has a fixed volume (V) of 100 ml and inflow–outflow rate (F) of 50 ml/min. By definition, *rate* is units of volume moving per unit time, but the kinetic behavior of the system also may be assessed in another sense. What *fraction* of the content of the pool is being replaced per unit time? In one minute this obviously is 50 ml/100 ml or 50%/min. Such fractional loss is known as a *rate constant*. The rate constant is 0.5 (or more explicitly 0.5/min or 0.5 min^{-1}). It will be assigned the symbol k. Thus in a pool of liquid undergoing volume flow,

$$k = F/V \qquad (1a)$$

Figure 1B is strictly comparable to A save that it represents mass rather than volume. Thus size (Q) is in mass units (here milligrams), and F now represents input–output rate in milligrams per minute. The rate constant now is fraction of 100 mg removed (and replaced) per minute,

$$k = F/Q \qquad (1b)$$

If multiple exits should exist (Figure 1C) each will have a separate k value, and the rate constant for the pool as a whole will be the *sum of all rate constants*.

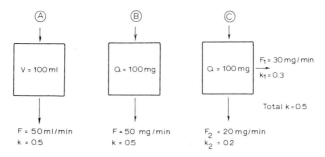

FIG. 1. Single pool systems wherein F is input–output rate, and k is the output rate constant. Model A is for liquid comprising volume, V, and B and C are for mass comprising weight, Q.

Tracer atoms and rate constant

4. One prime purpose of the tracer method is to calculate a flow rate when it cannot be measured directly. If the rate constant is known, this rate is calculable via a rearrangement of Eqs. (1a) and (1b). For mass,

$$F = kQ \qquad (2a)$$

or, if size is to be calculated,

$$Q = F/k \qquad (2b)$$

Consider that the purpose of adding tracer to the pool is to determine k. Tracer is assumed to mix with tracee almost instantaneously and remain continually mixed. Consequently, probability dictates that during any given time the chance of loss of a tracer atom at the site of output is the same as that for an unlabeled (tracee) atom. Therefore one may predict that loss of tracer should be 50%/min, and that the rate constant for tracer should be 0.5, as was the case for tracee atoms. But, as an experiment, add 1000 units of tracer to the pool. The initial concentration is 10 units per milliliter. At one minute the content of tracer will be 600 and its concentration 6. This is a 40% decline per minute rather than the 50% as predicted for accompanying tracee by Eq. (1a). The discrepancy arises because in one instance the fraction lost is the ratio of amount lost to a *fixed* amount of tracee during one minute, whereas for tracer the reference amount progressively declines. The true value for fraction lost per unit time can be approached arithmetically by making the time so short that the reference value undergoes minimal change. A general expression for estimation of such fractional rate of loss is

$$\text{estimate of } k \approx \frac{(q_0 - q)/q_0}{t} \tag{3}$$

The symbol q_0 is starting amount of tracer in the pool at zero time, and q is the amount observed later at time t. Direct measurement would give the following for expression (3):

t	q	k
0	1000	—
1	607	0.39
0.5	778	0.45
0.2	905	0.48
0.1	951	0.49

The approached limit of 0.5 is predicted directly by the calculus of Appendix I, which leads to the following equation for the time curve for quantity:[†]

$$q = q_0 e^{-kt} \qquad \text{or} \qquad q = De^{-kt} \tag{4a}$$

At the beginning the amount in the pool (q_0) is the whole dose (D). The symbol e is the base for natural logarithms. Equation (4a) is converted to one for *concentration* (c) simply by dividing by the constant volume (V) to give units of tracer per milliliter,

$$\frac{q}{V} = \frac{q_0}{V} e^{-kt} \qquad \text{of} \qquad c = c_0 e^{-kt} \tag{4b}$$

Likewise, if pool units are for mass, a division by weight of contained material (Q) converts total units present to units per milligram, i.e., specific activity (SA):

$$\frac{q}{Q} = \frac{q_0}{Q} e^{-kt} \qquad \text{or} \qquad SA = SA_0 e^{-kt} \tag{4c}$$

[†] A more formal notation would be

$$q(t) = q_0 e^{-kt}$$

The parenthetic t means "as a function of time." To keep clutter to a minimum, it will be omitted in this chapter when t is obviously the independent variable.

5. Equation (4b) is shown as a linear plot on the left of Figure 2. When plotted on a semilog scale, it follows a straight line (center and right). It may be shown that the slope of the line on the extreme right is k. (Note that Eq. (3) was an estimate of such slope.) Take the log to the base e of both sides of Eq. (4b). (This natural logarithm has the symbol ln.) Rearrangement then gives an expression in terms of k,

$$k = \frac{\ln c_0 - \ln c}{t} \tag{5a}$$

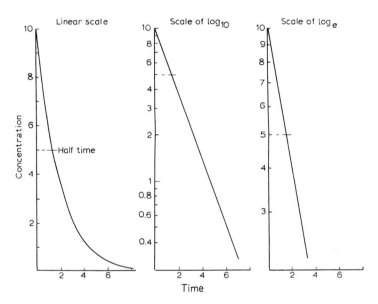

FIG. 2. The function, $c = 10e^{-0.5t}$, plotted on different scales as indicated.

This equation is for the first segment of the curve between zero time and time t, but the same relationship holds between any starting point t_1 when concentration is c_1, and a later point t_2 when concentration is c_2. Thus,

$$k = \frac{\ln c_1 - \ln c_2}{t_2 - t_1} \tag{5b}$$

Note that this is a two-point formula for slope of a line, but units on the ordinate are *natural logarithms* of c rather than c itself. It is the log scale in the two plots on the right which converts the curving line of the linear scale to a straight line with constant slope because distances on the ordinate are made proportional to logarithms. This is true regardless of the log base, but note that the slope is not the same in the two. Equations (5) apply only to the natural log. They are not applicable to $\log_{10}$. This exponential slope must not be confused with the slope of a curve based on actual values of q or c. For q (rather than $\ln q$), such slope at any instant is *quantity* lost per unit time rather than *fraction of existing quantity* lost. It continually diminishes toward zero slope at infinite time when ordinate values also approach zero (left-hand plot of Figure 2).   If this curve were obtained experimentally and c at 3 min were observed to be 2.23, Eq. (5a) would read

$$k = \frac{\ln 10 - \ln 2.23}{3} = \frac{2.303 - 0.802}{3} = 0.5$$

Ln N can be converted to units of $\log_{10} N$ by the relationship

$$\ln N = 2.303 \log_{10} N$$

A more convenient working formula is in terms of half time ($T_{1/2}$), which can be read directly from a plotted curve. Half time is the time required for c (or q) to decline by half anywhere on the curve, i.e., from 10 to 5, 0.2 to 0.1, etc. The formula is a modification of Eq. (5b) with c_2 equal to $c_1/2$ and $t_2 - t_1$ defined as $T_{1/2}$. With these substitutions and rearrangement,

$$k = \ln 2/T_{1/2} = 0.693/T_{1/2} \tag{5c}$$

In the present example, with $T_{1/2}$ being 1.39,

$$k = 0.693/1.39 = 0.5$$

For curves which decline rapidly, a more convenient formula employs time to complete one cycle, i.e., from 10 to 1, 0.9 to 0.09, etc. In Eq. (5b) substitute $c_1/10$ for c_2, so that $t_2 - t_1$ represents one cycle. Then

$$k = \frac{\ln 10}{\text{time for 1 cycle}} = \frac{2.303}{\text{time for 1 cycle}} \tag{5d}$$

In the present example,

$$k = 2.303/4.6 \text{ min} = 0.5$$

That the exponential slope is negative is indicated by the minus sign of the exponent in Eqs. (4).

VARIANTS OF BASIC EQUATIONS

6.   Rearrange Eq. (4a),

$$q/q_0 = e^{-kt} \tag{6a}$$

This is now a curve (a function) expressed as fraction of dose present versus time. Or, for SA observed as units of tracer per unit weight converted to fraction of dose per unit weight,

$$\frac{SA}{q_0} = \frac{SA_0}{q_0} e^{-kt} \tag{6b}$$

Still other forms with expressions (1) substituted for k are

$$q = q_0 e^{-(F/V)t} \quad \text{or} \quad q = q_0 e^{-(F/Q)t} \tag{6c}$$

A point on a curve may be taken at a time value beyond zero, e.g., at t_1, where the ordinate is q_1, and a value predicted for q_2 at time t_2. Equation (5b) may be expressed in exponential form as

$$q_2 = q_1 e^{-k(t_2 - t_1)} \tag{6d}$$

To construct a curve from any of the foregoing exponential equations, multiply k by successive values of t. This multiple is the exponent called X in most tables for e^{-X}.

INTERACTION OF RATE AND POOL SIZE

7. The plotted curves in Figure 3 are from the models as shown. The slopes of B and C are reduced by half when compared to A. Note that this is accomplished either by halving input–output rate (F) or doubling pool size (V). Equation (1a) obviously predicts this. The downward displacement of C as compared to B arises from the change in starting value of SA to 5 rather than 10. Figure 4 shows a simple tapwater analogy with volumes of beakers and flow rates corresponding to the values for those in Figure 1. If a dose of dye is added to each beaker, the fading of color with time is intuitively predicted to be retarded either by decreased flow rate or increased pool size. The effect of pool size on rapidity of tracer loss is an important consideration in a biologic setting. To be strictly avoided is the temptation to relate rapidity of tracer loss solely to the input–output rate of tracee.

POOL SIZE NOT KNOWN

8. Assume that an experiment yields the curve of Figure 3C but that, for practical reasons, sampling cannot begin until one minute from time of introducing tracer. Obviously, the straight line on semilog plot can easily be extrapolated back to zero time. It intersects at 5 on the ordinate, which is concentration predicted at zero time (c_0) if this had been measurable. The dose (q_0) placed at zero time is known to be 1000 units. Therefore, since $c_0 = q_0/V$, $5 = 1000/V$, and $V = 200$ ml. If units are milligrams for mass, $SA_0 = q_0/Q$, and $Q = 200$ mg. Although this approach has had wide usage for estimation of pool size in biologic systems, it may lead to gross error for the

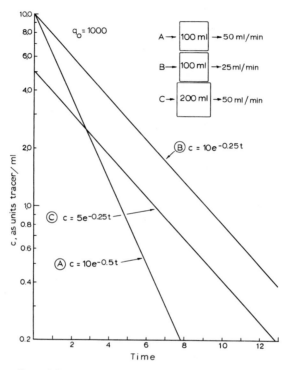

FIG. 3. The effect of flow rate and pool size on concentration curves for tracer.

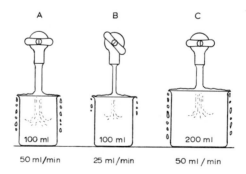

FIG. 4. Dye washout analogy.   Doubling the beaker size (C) has the same retarding effect on clearing of added dye as reducing flow rate to half (B).

simple reason that, as previously noted, pure single pools having no attachment to other pools are nonexistent in the animal body. Justification and limitations for lumping two or more pools of a composite system to treat the system as a single pool will be discussed later. The attractiveness of the maneuver of extrapolating for pool size rests in the power thereby vested in Eq. (2). Everything is solved by a simple SA curve. The slope gives k, the extrapolation gives Q, and the equation gives F.

Turnover

9. Turnover is a term frequently used in connection with a single pool whether it stands alone or is part of a multiple pool system. *Turnover rate* is simply the total steady-state quantity of tracee moving through (in or out of) the pool per unit time. In the present models it is F, kQ, or kV. If a pool should have multiple channels of influx and/or efflux, *whether connected to the outside or reversibly connected to other pools*, the turnover is the sum of the rates for *all* channels of exit, which is also the sum of rates for all channels of entrance. *Fractional turnover rate* is the same as the *fractional rate constant*, i.e., the *rate constant*. In the present single-pool models it is k. If multiple channels of output existed (including reversible output to other pools) the *overall* value for the pool would be the sum of *all* separate rate constants. Channels of input are not pertinent because a rate constant applies only to *removal*. *Turnover time* is the time required for that quantity which is exactly equal to pool size to move in and out. In model A of Figure 1 replacement of 100 ml at a rate of 50 ml/min requires $100/50 = 2$ min. Or if pool size were not known and the rate constant was known to be 0.5, the time to replace 100% of size V would be (100% of V)/(50% of V). This, expressed as fraction of total rather than percent of total, is $1/0.5$. Thus,

$$\text{turnover time} = V/F, \quad \text{or for mass,} \quad Q/F \tag{7}$$

Also,

$$\text{turnover time} = 1/k \tag{8}$$

Clearance

10. In 1928, Möller, McIntosh, and Van Slyke noted that if urine output was adequate, the excretion rate of urea was proportional to its concentration in blood. For example, with a blood urea nitrogen concentration of 0.16 mg/ml the excretion rate was 12 mg/min, and with a concentration of 0.20 mg/ml the excretion rate was 15 mg/min. Thus, $12/0.16 = 15/0.20$. Dividing either 12 by 0.16 or 15 by 0.20 gave 75, which was the number of milliliters to contain the quantity of urea nitrogen excreted per minute. The value of 75 ml/min was considered as volume nominally "cleared." Although even renal blood itself is never completely cleared, what is important is that a *constant fraction* of urea existing in any given volume of systemic blood is removed per unit time (and replaced in steady state). If half of the urea existing in any volume of blood were removed, 150 ml would be "half cleared" every minute to give the measured excretion rate of 15 mg/min. But the fraction removed not being known, the alternative is to express clearance as the volume nominally wholly cleared. It bears a fixed ratio to the volume which is fractionally cleared. Note that clearance is not a rate in the sense of quantity excreted per unit time. Rate can vary while clearance remains constant. Clearance is akin to a rate constant, as previously discussed. It sometimes is referred to as a clearance constant. This relationship may be appreciated intuitively because a rate constant represents a fixed fraction of total mass of material in a pool removed per unit time, and clearance is proportional to the fixed fraction of dissolved material which is removed from any volume per unit time (including total pool volume). Figure 5 is a model of a single pool in steady state containing

FIG. 5. A single pool to illustrate clearance calculations.

200 mg (Q) of a substance in 5000 ml volume (V). The material is excreted at a rate of 5 mg/min. The interconvertability of clearance and the rate constant of the pool with mass Q may be shown by the simple algebra which follows.

Clearance and rate constant

11. The formula for clearance (Cl) of unlabeled material, having concentration in water c, is

$$Cl = \frac{F}{c} = \frac{5 \text{ mg/min}}{0.04 \text{ mg/ml}} = 125 \text{ ml/min} \tag{9}$$

Rearranging in terms of excretion rate and substituting Q/V for c,

$$F = Cl \frac{Q}{V} = 125 \frac{200}{5000} = 5 \text{ mg/min} \tag{10}$$

Now assume that a dose of radioactive tracer material is placed in the pool and a curve of SA, i.e., q/Q, versus time is obtained. From Eq. (1b) the slope is predicted to be 0.025. Arranged as in Eq. (2a),

$$F = kQ = (0.025)(200) = 5 \text{ mg/min} \tag{11}$$

Equating expressions (10) and (11),

$$Cl \ Q/V = kQ$$

Dividing both sides by Q and rearranging,

$$Cl = kV \quad \text{or} \quad k = Cl/V \tag{12}$$

Thus, the rate constant and clearance are directly proportional to each other, but their absolute values will vary with the volume in which the unlabeled substance is contained. If one of the two is measured the volume of distribution must be known to determine the other. As was true for rate constants, if two or more sites of clearance exist (e.g., two or more organs clearing blood) the overall clearance is the sum of all.

Clearance of tracer as such

12. Because natural and tracer atoms are handled alike, the clearance of tracer from water has the same value as clearance of associated tracee atoms from water. A theoretical limitation of tracer for the purpose is that its concentration in water declines with time after a single dose. Nevertheless, a satisfactory approximation is possible if a short interval of time is used and a simple mean is taken for tracer concentration in water. For example in the case of the thyroid gland, where ^{127}I movement to the gland is not measurable, ^{131}I may be used instead. Over a two-hour period, the uptake of ^{131}I in the gland divided by 120 min gives an approximation of a mean rate of uptake per minute. This is comparable to F in Eq. (9). During this same period, the concentration of ^{131}I in blood is averaged to serve as c in the equation. Measurements should be confined to the first 2 or 3 hours while all ^{131}I in blood is donor iodide, and appreciable hormonal ^{131}I has not yet begun to enter blood from the gland. For complex systems see also Chapters 6 and 9.

Biologic implications of clearance principle

13. Because clearance (or a rate constant) tends in fact to be *constant* for a normally functioning organ, it serves as an intrinsic homeostatic mechanism. In the case of kidney, for example, if dietary protein is increased, and blood urea, therefore, tends to rise, the constancy of clearance causes output rate of urea to increase automatically, which in turn tends to lower blood concentration. The reverse is true if dietary protein is reduced. The rate of urea excretion obviously is not a measure of the integrity of renal function. On the other hand, renal clearance is an appropriate index of such integrity. Reduced clearance means a reduced fraction removed from the blood encountered by renal tissue. Clearance, in a sense, reflects the intensity of cell function as divorced from load. In the normal subject, thyroid clearance of iodide from plasma is close to 15 ml/min. In hyperthyroidism it averages about 200. Although thyroid clearance is independent of plasma iodide concentration *within certain limits*, an excessive load can saturate the extraction mechanism. After pharmacologic doses of iodide, the thyroid is unable to remove its usual fractional quota of iodide from plasma (including the fraction of associated tracer) because this embodies too many iodine atoms for it to handle. Clearance no longer is unrelated to blood concentration. This saturation effect would reduce clearance even if iodide had no direct inhibiting effect on hormone synthesis.

Estimation of Size and Turnover Rate of Two Pools Lumped as if One

SIZE OF THE SYSTEM

14. In paragraph 8, the estimate of size of a single pool by extrapolation of the SA curve to zero time was shown to be straightforward and theoretically accurate. In a system consisting of two or more pools, such an extrapolation will not give the size of the complex as a whole except under certain special conditions. One such condition is a closed system as shown in Figure 6A. If

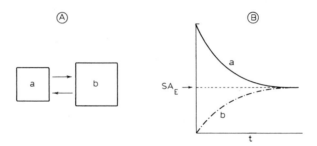

FIG. 6. Estimation of the size of a closed system by extrapolation. SA_E is the ordinate value at equilibrium.

tracer is placed in pool a at zero time the SA in a falls and that of b rises to a stable permanent equilibrium represented by the horizontal line SA_E in Figure 6B. Dose of tracer divided by the zero intercept of this line obviously gives $Q_a + Q_b$, the mass of unlabeled material in the entire system. From this example of a closed complex system and that of an open single pool, the temptation may be strong to apply the method to any *open complex system.* Most in vivo biologic systems are in this category. Figure 7A is a simple example of an open double pool system with pool a interchanging with pool b while inflow and outflow proceed as shown. Figure 7B is a plot of SA for pool a on a semilog scale. In the early days of tracer usage a common practice was to extrapolate the straight tail of the curve of the labeled pool (dashed line) to predict the hypothetical concentration at zero time in an *instantaneously* mixed system. The inflected portion of the curve prior to one hour was

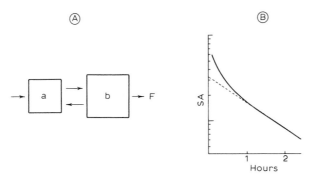

FIG. 7. A two-pool open reversible system and a semilog plot of the SA curve for pool *a*. The dashed line is an extrapolation of the tail.

considered to be the mixing phase between the two pools (not actually instantaneous), whereas the straight tail was attributed to a state of equilibrium similar to that in the closed system of Figure 6A. And, as in the closed system, the SA value at the zero intercept was divided into the value for dose to give composite size of the two pools together. The errors attending such an approach now will be examined in detail.

15. The error in treating the model of Figure 7A in the same way as that of Figure 6A arises from the fact that when interchange between the pools is not extremely rapid the process of intermixing never brings about a *sustained equal concentration* of tracer throughout the system. For the open pool model (Figure 7A), the SA curves for pools *a* and *b* will look like those of Figure 8. The sharp decline of activity in pool *a* during the first hour reflects the dominant influence of the rapid net movement of tracer into pool *b* at this early point in time. SA in pool *b* rises during the first hour, and only for a mathematical instant at its peak does it equal that of pool *a*. Pool *a*, being in

FIG. 8. A semilog plot of SA curves for each pool in an open two-pool reversible system such as in Figure 7A.

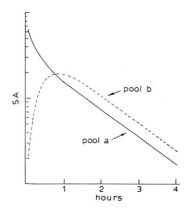

position of direct access to inflowing diluent, will from this point onward, have a SA less than that of previously enriched pool *b*, and the tails of the curves will become parallel because net fractional loss now is comparable in the two pools. Obviously, an extrapolation of the curve for *a* gives a different zero intercept than that for *b*. Neither value is that for hypothetical SA at t_0 with instantaneous mixing. The extrapolated value for *a* is too low. Dose divided by hypothetical SA_0 will overestimate the size of the system.

EXAMPLES OF EFFECTS OF LUMPING

16. A representative model to which symbols for flow rate and capacity now are introduced is shown in Figure 9. The pools are considered to consist of identical material undergoing interchange between *a* and *b* as by diffusion. At zero time, tracer is introduced into pool *a*. Mixing is assumed near instantaneous and continuous only *within* each pool and not between the two. The dose of tracer will be called 1 for simplicity. Pool *b* is reversibly connected with pool *a* and it serves as a discharge route at a rate of outflow F_{ob} equal to inflow to pool *a* (F_{ao}). F_{ba} is the rate of flow *to b from a*, and F_{ab} is the reverse.

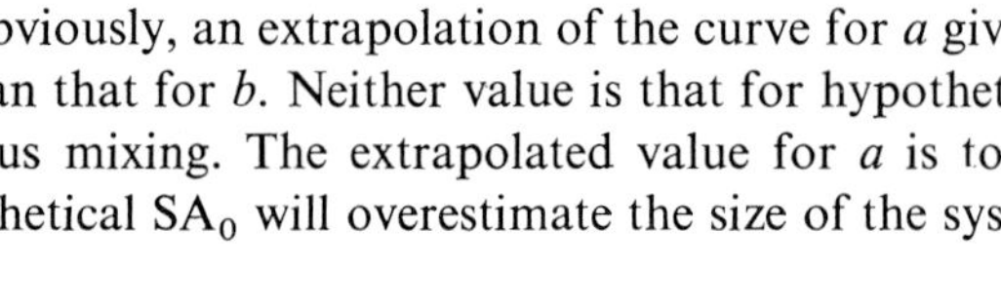

a   b
                (0.03)
                F_{ba}
F_{ao} → $Q_a = 1$ mg   ⟶ $Q_b = 2$ mg → $F_{ob} = 0.02$ mg/min
                F_{ab}
                (0.01)

FIG. 9. An explicit two-pool open reversible system. Values are employed in calculations shown in the text.

The subscript notation, *ba*, denoting *to b* from *a* (rather than *from b* to *a*) is not found universally in published literature, however it is now the preferred convention (Brownell *et al.*, 1968). The amounts of unlabeled material in pools *a* and *b* (Q_a and Q_b) are 1.0 mg and 2.0 mg, respectively. The overall inflow–outflow rate is 0.02 mg/min, F_{ba} is 0.03 mg/min, and F_{ab} is 0.01 mg/min.

17. Figure 10 shows what the curves of specific activity look like in this system (solid lines). Again, note that an equal concentration of tracer in the two pools exists only at one point, i.e., the peak SA for pool *b*. After about 200 min the pools are equilibrated only in the sense that their curves become effectively parallel. The error in calculating the composite size of pools *a* and *b* is evident if the extrapolated zero intercept of pool *a* is used for the calculation. The value 0.241 divided into dose (1.0) gives 4.15 mg, as compared to the true value of 3.0. The slope of the terminal portion of the SA curves can be shown graphically to be 0.0081. This multiplied by the erroneous pool size gives 0.034 mg/min for supposed rate of outflow instead of the true value of 0.02. The dot–dash line is what the actual intercept and slope would look like if the two pools mixed instantaneously, i.e., were in fact a single pool of size 3.0 mg. The SA at zero time would be dose/mg in pool = 1.0/3.0 = 0.333. The slope would be 0.02/3.0 = 0.0067 (Eq. (1b)). Hence, the equation for the line is SA = $0.333e^{-0.0067t}$. Note that the slope (in addition to zero intercept) differs from that of the observed curves. It should now be clear that one's ability to draw a straight line through terminal points of observed data plotted on semilog paper is not, per se, justification for treating the system as a single pool.

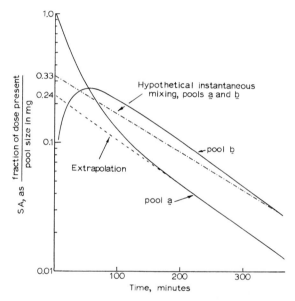

FIG. 10. Comparison of actual SA curves (for the two pools of Figure 9) with the curves for hypothetical instantaneous interpool mixing (semilog plot).

18. Plots of curves of specific activity for various other rates of interchange between pools a and b are shown in Figure 11. Some of the actual exchange rates are given in Table I. The family of curves lies between two limiting simple exponentials (straight lines). The dot–dash line of instant interchange (uppermost after 120 min) is the same as that of the hypothetical curve in Figure 10. The two pools behave as though they did not exist separately but constituted one large pool of size 3.0. Curve G falls very close to this line. The other dashed line at the bottom represents the hypothetical situation where no reflux to a from b exists. Its equation is $c = 1.0e^{-0.02t}$, because F_{ba}, now a simple nonreversible exit, $= F_{ob} = 0.02$, and $k = 0.02/1$ (Eq. (1b)). This line represents the limit at which pool *a* is an isolated single pool oblivious of the existence of pool *b*. Note in Table I that although the error in estimate of Q_a plus Q_b is small when reflux (F_{ab}) is very rapid in relation to rate of outflow (F_{ob}), at the other extreme the estimate is very wide of the mark. In the region between curves A and B data points defining an experimental curve might be so uncertain that a decision could not be made whether the line curves or is straight. The points for curve A certainly could be judged to follow a straight line. The best visual fit between 0 and 300 min would be consistent with a straight line very close to that for a single nonreversible pool *a*.

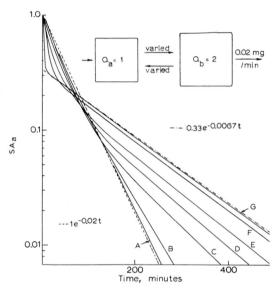

FIG. 11. Effect on SA curve of pool *a* when rate of interchange between pools is varied. See text and Table I.

19. Worthy of emphasis is the fact that the terminal segments of complex exponential SA curves obtained from the various compartments of any reversibly interconnected system of pools, regardless of the number, all approach the same slope. They differ only in their vertical position. If interchange at *any* interpool connection is very slow in relation to the rate of external exit the establishment of this final common slope will be long delayed in all pools. For example, curve *A* in Figure 11 does not attain the final slope until a time far beyond that included in the figure. By methods to be presented in Chapter 2, the extrapolated zero intercept of the final segment is predicted to be 0.0097, and its slope 0.00995. For this curve $F_{ba} = 0.0201$, and $F_{ab} = 0.0001$.

TABLE I

VALUES FOR FIGURE 11

F_{ba}	F_{ab}	Zero intercept (terminal slope)	Estimate of $Q_a + Q_b$ by extrapolation (mg)	Percent error in size of $Q_a + Q_b$ (%)	Terminal slope	Estimate of rate F_{ob} (slope × estimate $Q_a + Q_b$) (mg/min)	Percent error in estimate of F_{ob} (%)	Curve of Fig. 11
0.021	0.001	0.0806	12.42	313	0.0096	0.119	493	B
0.030	0.010	0.241	4.15	38	0.0081	0.034	70	D
0.040	0.020	0.276	3.62	21	0.0076	0.028	38	E
0.100	0.080	0.316	3.17	6	0.0070	0.022	9	F
0.600	0.580	0.331	3.02	1	0.0067	0.0202	1	G

20. Another feature of a pool system which affects errors in estimate of total size by extrapolation of the tail is the relative size of the secondary pool(s). In the example just given, the secondary pool was twice as large as the primary labeled pool. The larger the secondary pool, the greater the error. This follows from the fact that a great deal of tracer is held up when the secondary pool is large. With any given rate of interflow, relatively less tracer returns to the primary pool during the early phase of observation because tracer in the refluxed material is in relatively dilute concentration. Consequently, the initial steep downslope of the curve is more prolonged than when the secondary pool is small. In a model where the size of pool b is allowed to vary, F_{ob} is fixed at 0.02, and F_{ba} and F_{ab} are 0.025 and 0.005, respectively, the effect of the capacity of pool b is shown in Figure 12. Note that the intercept of the tail is 0.002 when pool b is 100 mg (Q_b). This would give a calculated joint size of 500 when actually the total size is 101. The terminal slope of 0.0002 would give 0.1 mg/min for apparent turnover

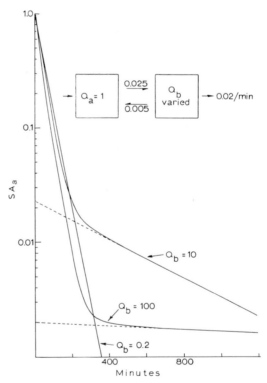

FIG. 12. Effect on SA curve of pool a when size of secondary pool b is varied. Rates in the model are milligrams per minute.

rate of the system. On the other hand, if pool b is only 0.2 mg in size the decay curve is visually indistinguishable from a straight line having an intercept at 0.95, which means that joint pool size would calculate to be 1.05, a rather slight underestimate of the true value of 1.20. The estimated slope of the apparent straight line would be 0.0192 and calculated turnover rate 0.0202 mg/min.

21. Another manipulation of interest is that of reducing outflow (F_{ob}) to vanishingly small values (Figure 13). Pool sizes will remain at 1.0 and 2.0,

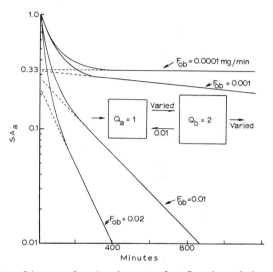

FIG. 13. Effect on SA curve of pool *a* when rate of net flow through the system is varied.

respectively. F_{ab} will be fixed at 0.01 mg/min and F_{ob} allowed to vary. The lowermost curve of Figure 13 is the same as D of Figure 11 and the curve for pool *a* of Figure 10. With an outflow rate one-tenth that of reflux rate ($F_{ob} = 0.001$) the joint pool capacity is calculated at 3.08, an error of 2.7%. When output rate equals reflux rate ($F_{ob} = 0.01$) these respective values are 3.62 and 21%.

22. The following generalizations now can be made: In physiologic systems involving irreversible disposal of a compound by excretion or chemical conversion, a pure single-pool system is not to be expected. Actually, an animal body is composed of a near-countless number of physical and chemical pools. Although for practical purposes to perform kinetic analysis it is obligatory to consolidate certain moieties of the system into one-pool equivalents, such simplification has limitations. When a primary pool and an attending system of interconnecting pools are treated as a single compartment for purposes of estimating the joint pool size and the rate of disposal from the system as a whole by simple extrapolation of the terminal slope, the error may be very large. Such error is increased in proportion to the relative slowness of interchange between the primary labeled pool and secondary pools with which the exchange occurs, and in proportion to the size of any of these secondary pools. Although intuitive judgment can be helpful, as when one feels justified in accepting an extrapolation of a long terminal segment preceded by a very sharp and short initial segment of rapid decline, a knowledge of the nature and size of the separate pools making up a complex system, as provided by independent information, is most helpful. If this information is available the mathematical analysis more appropriately might be that for a multicompartmental system of two or three pools rather than of an unnecessarily simplified single-pool model. In Chapter 2, paragraph 15, a two-pool model with separate channels of both *inflow and outflow* to *each* pool will be demonstrated, where the extrapolation of terminal slope does indeed give correct size of the system and turnover rate. But the chance of encountering such a unique model with appropriate ratios of flow rates and pool sizes in a biologic setting is remote.

23.   Finally, another source of potentially serious error in "working blind" with only a curve of specific activity is to mistakenly assume that equilibrium of tracer occurring during an initial sharp downslope is a physical mixing process, if in fact it is a reversible chemical reaction wherein carbon atoms, for example, rather than unchanged molecules, are shuttling back and forth between the compound known to exist in the first pool and another different compound in a second pool. With such a chemical cycling of carbon atoms, the joint pool size calculated by extrapolation of the terminal segment of a curve of ^{14}C specific activity would be augmented by carbon of an entirely different compound in the secondary pool. Such a mistaken identity of a secondary pool might introduce the most serious error of all.

Excretion of Drugs

24.   The preceding discussion pertained to body compartments containing natural constituents which, upon leaving a pool, are replaced by like material. The tracer was a counterpart to natural constituents. In the case of a drug or a nonregenerated compound whether radioactive or not, its loss from blood, either by degradation or excretion, is usually subject to the same exponential process as has been described for tracer atoms. All of the limitations previously recounted apply here also, whether for the estimation of volume of distribution, or for the rate constant of loss from the body. Fortunately, however, many foreign compounds such as drugs have a clearance rate from the body which is much slower than for intermixing within compartments. The errors, therefore, may not be great. The curve for blood content might resemble curves F or G of Figure 11.

COMPARTMENT ANALYSIS: TWO-POOL OPEN SYSTEMS

Multiple-Compartment Analysis

1. In the latter half of Chapter 1, the errors inherent in treating a two-pool system as though it were one pool were documented with illustrative examples. Chapter 2 will describe how, from curves of tracer content of the labeled pool or of the secondary pool, a definitive evaluation is made of rate constants and rate of interflow and outflow in the two-pool open system, and how the ratio of size of one pool to the other is calculated. Also to be considered is the limited usefulness of a curve of *specific activity* (instead of tracer content) when information is restricted to such a curve obtained from the secondary pool. Methods for multiple compartment analysis have been given by Skinner *et al.* (1959), Lewallen *et al.* (1959), and Gurpide *et al.* (1964). For many early references, see also Robertson's review (1957) and Sheppard's book (1962). Compartment analysis follows from an assumption that transfer of tracer can be described by exponential functions. The simple exponential function attending a one-pool system has been described in Chapter 1. As the number of compartments increases, the computation not only becomes more complex but the components in the curve become so numerous that graphic analysis may be impractical. These problems will be discussed in Chapter 4.

Simplification may be possible by lumping several rapidly interchanging pools into one common pool. Thus, a ten-pool system may have a subsystem *A*, of six pools undergoing interchange very rapidly with each other as compared to the rate of interchange of this aggregate group with another subsystem, *B*, of four pools which also have very rapid interchange with each other. This overall ten-pool conglomerate might be simplified to a two-pool system consisting effectively of (lumped) compartment *A* and (lumped) compartment *B*. The pitfalls of injudicious lumping of compartments having relatively slow interchange or markedly disparate size are described in Chapter 1.

Analysis of an Interchanging System

Pools, SA, Quantity of Tracer

2.   Figure 14 will serve as an illustrative model for the calculations to be performed. The system could be a physical one wherein pools *a* and *b* represented spacial compartments undergoing molecular interchange while net movement was to the right. But in the present example it will be considered

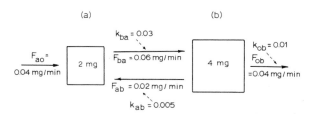

FIG. 14. A model to illustrate compartment analysis. Rate constants (*k*) are shown along with transport rates for unlabeled material (*F*).

as a chemical system embodying carbon atoms. Pool *a* is carbon in molecular species *a*. Species *a* and *b* are chemically interconvertible. Carbon atoms interchange as a part of this process. Conjointly in this reaction other atoms such as hydrogen and oxygen also may participate in the interchange and may enter and leave the system, however these movements are not shown because the tracer will be ^{14}C, and movement of tracer carbon will tell nothing about the movement of noncarbon atoms. Pools *a* and *b* represent separate molecular species in which carbon atoms reside. Pool sizes (Q_a and Q_b) are in terms of milligrams of *carbon* residing therein, and flow rates (*F*) are milligrams of carbon per minute. Pool *a* contains 2 mg of carbon. The quantity of *compound* is not specified. If it should contain 50% carbon by weight, the quantity of compound therein would be 4 mg. If compound b should contain 20% carbon, the total weight of molecular species would be 20 mg and its rate of output (F_{ob}) would be 0.2 mg/min. But for primary calculations the material in a pool and the material moving is carbon. Likewise the specific activity (SA) is in terms of tracer per unit weight of *carbon*. Added tracer will ordinarily be of such small mass that it may be neglected as a contributor to weight.

3. Pool a will be labeled, i.e., it will receive species a containing ^{14}C uniformly distributed as label. The dose will be given abruptly at zero time. The number of units of tracer existing in pool a at any time will be called q_a and that introduced at zero time will be referred to as q_{a0}. For convenience the quantity of tracer existing at any given instant is best designated as fraction of dose, i.e., q_a/q_{a0}, rather than as "counts" or microcuries. This not only will compensate for possible differences in dose from one experiment to another, it will also fit directly in equations to follow which are best defined in terms of fraction of dose. Thus, this dose, for working purposes becomes 1 (or alternatively 100%) and SA becomes fraction of dose (or alternately percent of dose) per unit weight of unlabeled atoms (carbon in this instance). If first measured as *radioactivity* per milligram, SA is converted to *fraction of dose* per milligram simply by dividing such SA by the total activity in the dose. And because activity per milligram is equivalent to the ratio of total activity in the pool (q_a) to pool size (Q_a), the following arrangements are equalities:

SA (adjusted to fraction of dose/mg)

$$= \frac{\text{Activity/mg}}{\text{Activity in dose}} = \frac{q_a/Q_a}{q_{a0}} = \frac{q_a}{Q_a q_{a0}} = \frac{q_a/q_{a0}}{Q_a}$$

Likewise, the number of units of tracer in pool b at any time is q_b, the fraction of dose is q_b/q_{a0}, and SA expressed as fraction of dose per milligram is $(q_b/Q_b)/q_{a0}$ etc. Digressing from the specific model at hand, a convenient correction for SA measured in any compound of a living animal should be mentioned. All else being equal, 200 gm of total rat, for instance, will dilute a given dose of tracer twice as much in all participating compartments as will 100 gm of total rat. Thus a rat of explicit size should be chosen as standard. Should this be 100 gm, the SA of the atoms in all compounds of a rat weighing 200 gm must be multiplied by 2. The correction factor, therefore, is

rat weight in grams/100 gm

This ratio multiplied by one of the foregoing expressions predicts SA adjusted to represent fraction of dose per mg if all rats weighed 100 gm. The value of

Q_a is that for a 100-gm rat. Such an adjustment permits averaging sets of SA values obtained at comparable times when doses and rat weights are not identical.

OBSERVE SA OF POOL a

Graphic curve analysis

4. Assume that the only quantitative information available is a specific activity curve from pool a of Figure 14 beginning at 10 minutes and continuing for 5 to 6 hours (Figure 15). The scale of SA is fraction of dose per milligram carbon. The actual dose, for example, might have been 150,000 tracer units, and the 10 minute reading 55,500 units/mg carbon. The fraction of dose per

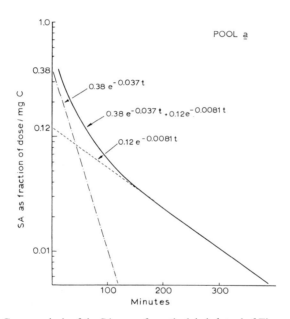

FIG. 15. Curve analysis of the SA curve from the labeled pool of Figure 14 (pool a).

milligram C at 10 min would be $55,500/150,000 = 0.37$, and so on. Thus, any point on the curve is SA_a/q_{a0}, where SA_a is tracer units per mg, and q_{a0} is the dose placed in pool a at zero time. A working assumption will be that this curve represents an exponential function, and that its failure to follow a single straight line in its entirety when plotted on semilogarithmic paper identifies it as a complex function. This means that the equation for the line embodies more than one exponential term. When two or more exponential terms are added together the resulting line is not initially straight. By a graphical approach the line may be dissected to yield the separate terms of the equation.

The process is sometimes called "curve peeling" or graphical analysis. It consists of a series of subtractions (in an equation with two terms only one subtraction). Note that the tail of the curve beyond 200 min plots as a straight line. In a practical sense, this segment represents one of the terms in pure form. It represents the term with the least steep slope, i.e., the term with the smallest exponent. The other term although persisting to infinity in a mathematical sense, contributes insignificantly beyond 200 min, because it declines relatively rapidly to comparatively low values by virtue of its steeper slope. By a method given in Chapter 1, paragraph 5, the tail of this curve is determined to have a slope of -0.0081. When extrapolated to zero time (Figure 15, dashed line) the intercept on the ordinate at 0.12 represents the coefficient for this term. This defines the "late component" (normally written last in the series of terms) as $0.12e^{-0.0081t}$. To find the early rapid component a series of values along the extrapolated dashed line at specific points in time are subtracted arithmetically from the corresponding values of the original composite curve. For example, at ten minutes the subtraction is $0.37 - 0.11 = 0.26$. Additional subtractions up to 150 min define points on a line which extrapolates to 0.38 at t_0 (dot–dash line) and has a slope of -0.037. This term therefore is $0.38e^{-0.037t}$. The total curve for a two-pool interchanging system will have just two components. Their numerical summation defines the time curve of specific activity for pool a which here is expressed as *fraction of dose per mg carbon*

$$SA_a/q_{a0} = 0.38e^{-0.037t} + 0.12e^{-0.0081t} \tag{1}$$

Conversion of SA to quantity of tracer

5. Quantity of tracer in pool a (q_a) obviously will be given by multiplying pool size in milligrams (Q_a) by specific activity. Pool size can be estimated by extrapolation, but unlike a single pool the primary curve cannot be drawn accurately to the ordinate by visual estimate because it is curving. The intercepts of the two straight lines must be added together. In this case the sum is 0.5. Because units of SA are in fraction of dose, the normalized dose becomes 1. This (being present at zero time), when divided by SA of pool a at zero time (SA_{a0}) gives milligrams in pool a,

$$Q_a = \frac{\text{dose}}{SA_{a0} \text{ (as fraction of dose/mg)}} = \frac{1}{0.5} = 2 \text{ mg}$$

The complete time curve for quantity as fraction of dose present in pool a can be drawn by multiplying all points on Eq. (1) by 2 mg. The equation then becomes†

$$q_a/q_{a0} = 0.76e^{-0.037t} + 0.24e^{-0.0081t} \tag{2a}$$

† If more exact values are desired for more accurate calculations this equation is
$$q_a/q_{a0} = 0.7611e^{-0.03686t} + 0.2389e^{-0.00814t}$$

This curve is shown in Figure 17B (along with that for pool b). This normalized equation for quantity expressed as fraction of dose is important because the coefficients are explicitly tailored for the compartment analysis to follow. The first coefficient in order (not necessarily always the larger) will be assigned the symbol H_1 and the second H_2. The slope of the first term will be called g_1 and of the second term g_2. The slope having the larger value is always called g_1, i.e., it is in the first term. The general normalized expression is

$$q_a/q_{a0} = H_1 e^{-g_1 t} + H_2 e^{-g_2 t} \tag{2b}$$

To convert the curve into one expressed in terms of numerical units of tracer rather than fraction of dose simply rearrange

$$q_a = q_{a0}(H_1 e^{-g_1 t} + H_2 e^{-g_2 t}) \tag{2c}$$

Alternate routes to the normalized equation

6.   Note that the coefficients (intercepts) in Eq. (2a) add to unity. This must be so because the whole dose (1) is present in the pool at zero time, and also because mathematically at zero time $e^{-g_1 t}$ and $e^{-g_2 t}$ each become 1 leaving simply $H_1 + H_2$ on the right. In essence these two coefficients are the fractions which, when added together, define the starting point of the curve. Expressed as percentages H_1 is 76% of the starting point and H_2 is 24%. This means that normalized intercepts can be calculated directly for any curve expressed in any units. Suppose, for example, the dose had been 150,000 tracer units. Then for SA as *counts* per milligram rather than fraction of dose per milligram,

$$SA_a = 57,000 e^{-0.037 t} + 18,000 e^{-0.0081 t}$$

$$H_1 = \frac{57,000}{57,000 + 18,000} = 0.76$$

$$H_2 = \frac{18,000}{57,000 + 18,000} = 0.24$$

Algebraically this identity is predictable from Eq. (2b). Changing to SA does not change the ratio,

$$\frac{q_a}{q_{a0}} = \frac{q_a/Q_a}{q_{a0}/Q_a} = \frac{SA_a}{SA_{a0}}$$

The same principle holds for a radiation detector looking at pool a and recording arbitrary units of count rate (R_a). Then,

$$R_a/R_{a0} = H_1 e^{-g_1 t} + H_2 e^{-g_2 t} \tag{2d}$$

In its most general form, the normalized equation for any curve having intercepts I_1 and I_2:

$$f(t) = \frac{I_1}{I_1 + I_2} e^{-g_1 t} + \frac{I_2}{I_1 + I_2} e^{-g_2 t} \tag{2e}$$

SAMPLING POOL b

Graphic curve analysis

7. Serial sampling of pool b (Figure 14) will yield the SA curve shown as a solid line in Figure 16. The relatively straight terminal portion of the curve has the same slope as that for the curve from pool a, i.e., -0.0081. It extrapolates to a zero intercept of 0.26. Subtraction of the extrapolated portion of this line from the early portion of the observed curve is performed as before but now the differences are negative because the extrapolated segment lies above the original curve. For example, at 50 min, the value on the extrapolated segment is 0.17 and that on the curve is 0.13. The subtraction is $0.13 - 0.17 = -0.04$. Therefore for the dot–dash line generated in this manner, the values on the ordinate are negative. The slope is -0.037 as was the case for the rapid component of the curve for pool a, and the zero intercept is -0.26. As fraction of dose per milligram carbon the total curve is

$$SA_b/q_{a0} = -0.26e^{-0.037t} + 0.26e^{-0.0081t} \tag{3a}$$

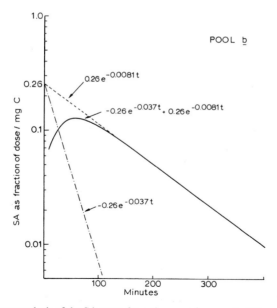

FIG. 16. Curve analysis of the SA curve from the secondary pool of Figure 14 (pool b).

Note that at zero time when the exponential portions of the terms are both unity and the two coefficients being of opposite sign add to zero, the value of the function is zero. No tracer as yet exists in pool b. The negative first term with the larger exponent serves to generate the initial upslope. Later on it contributes insignificantly and the second term yields a terminal tail with positive intercept. Curves for SA in both pool a and pool b are shown together in Figure 17A.

Conversion to quantity of tracer

8.   Unlike the equation for SA of pool a Eq. (3a) cannot be converted to one for quantity by any graphical or mathematical manipulation. No value for instantaneous concentration exists at zero time because the tracer content is zero at this time. When pool b *alone* is sampled for SA, its size, i.e., milligrams of carbon (Q_b), must be independently known in order to convert the curve to one for quantity. Multiplying Eq. (3a) by Q_b (known to be 4) gives

$$4(SA_b/q_{a0}) = 4(-0.26e^{-0.037t} + 0.26e^{-0.0081t})$$

which is quantity in terms of fraction of the dose placed in pool a

$$q_b/q_{a0} = -1.04e^{-0.037t} + 1.04e^{-0.0081t} \qquad (3b)$$

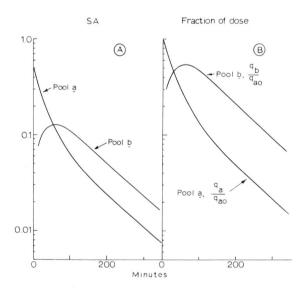

FIG. 17. Comparison of curves for SA (fraction of dose per mg C) and content of tracer (fraction of dose) in the two pools of Figure 14.

The curve, along with a corresponding one for pool a, is shown in Figure 17B. Equation (3b) is the completely normalized expression for *fraction of dose* existing in pool b. It will be assigned general symbols K_1 and K_2 for the coefficients (intercepts) as follows,

$$q_b/q_{a0} = K_1 e^{-g_1 t} + K_2 e^{-g_2 t} \tag{3c}$$

These specific coefficients will be employed in compartment analysis. Because their sum is zero they always are equal and of opposite sign, with K_1 being the negative member of the pair. Note that although in the primary labeled pool the ratios q_a/q_{a0}, SA_a/SA_{a0}, and R_a/R_{a0} were all equal and were associated with fractional coefficients H, the ratio q_b/q_{a0} has no such direct equivalents in terms of SA or detector readings obtainable from pool b *alone*. If matched detectors of identical geometry were viewing pools a and b simultaneously, only then would an equivalent ratio of q_b/q_{a0} be obtainable. With the readings versus time over pool b called $R_b(t)$, and that over pool a at zero time called R_{a0}:

$$R_b(t)/R_{a0} = q_b(t)/q_{a0}$$

Rearrangement to give quantity in terms of numerical counts or μCi, etc., yields

$$q_b = q_{a0}(K_1 e^{-g_1 t} + K_2 e^{-g_2 t}) \tag{3d}$$

Identity of slopes

9.   The exact match of counterpart slopes (g_1 and g_2) for the curves from the two pools is characteristic for curves generated by kinetic events in *reversibly* connected pools. If the system consisted of a three-pool mutually reversible system, three terms would appear in the equation for each pool, and the counterpart slopes would be identical in all three. In other words the number of components in each curve, or terms in the equation for each curve is the same as the number of mutually interconnected pools, and the successive slopes are identical no matter which pool is sampled. Only the coefficients vary. If one of the pools should have an inlet *from* but *no reverse* connection *to* the remainder of the system its curve still would contain components with slopes characteristic of the other pools which feed into it (plus another self-contributed component), however, it would not contribute its component to the curves, or terms to the equations, of the other preceding pools. This will be illustrated for a two-pool system in paragraph 17. An exception to the rule that the number of nonidentical slopes matches the number of pools is the unique steady-state one-way model where pools are the same size (paragraph 21).

Solution by Sampling Labeled Pool *a*

RATE CONSTANTS

10. Before illustrating calculations for the model of Figure 14 the intent of the process should be made clear. Neither of the two slopes, g_1 and g_2, is directly equivalent numerically to any rate constant of interflow or outflow. These rate constants are k_{ba}, k_{ab}, k_{ob}, k_{aa}, and k_{bb}. The latter two are the overall turnover rate constants for pools *a* and *b*, respectively. Each represents the respective sum of all rate constants of efflux from the particular pool. For pool *a*, $k_{aa} = k_{ba}$; for pool *b*, $k_{bb} = k_{ab} + k_{ob}$. Each rate constant is related mathematically to the entire group of parameters in the normalized Eq. (2b), namely g_1, g_2, H_1 and H_2. The calculation is directed toward evaluating the rate constants via equations which embody these four parameters as contained in Eq. (2a). Three expressions will be used for the solution: They are from Appendix II, Eqs. (13), (14), and (18a). In a two-pool system, the general equations are shortened by making all rate constants, slopes, and intercepts relating to the nonexistent third pool equal to zero. The following abbreviated expressions then emerge:

$$k_{aa} = H_1 g_1 + H_2 g_2 \tag{4}$$

$$k_{aa} + k_{bb} = g_1 + g_2 \tag{5}$$

$$k_{aa} k_{bb} - k_{ab} k_{ba} = g_1 g_2 \tag{6}$$

Slopes and intercepts obtained via curve peeling are substituted in Eq. (4),

$$k_{aa} = (0.76)(0.037) + (0.24)(0.0081) = 0.030$$

Substituting 0.03 in Eq. (5) and rearranging,

$$k_{bb} = 0.037 + 0.0081 - 0.03 = 0.015$$

Because k_{ba} is the sole exit from pool *a*,

$$k_{ba} = k_{aa} = 0.03$$

Substituting k_{aa}, k_{bb}, and k_{ba} in Eq. (6) and rearranging,

$$k_{ab} = \frac{(0.03)(0.015) - (0.037)(0.0081)}{0.03} = 0.005$$

From the model,

$$k_{ob} = k_{bb} - k_{ab} = 0.015 - 0.005 = 0.01$$

FLOW RATES AND SIZE OF POOL *b*

11. To this point all rate constants have been evaluated solely by the co-efficients and slopes of Eq. (2a). To determine the size of pool *b* (Q_b) and the flow rates (F) in milligrams per minute, the size of pool *a* (Q_a) must be introduced. It was previously calculated to be 2 mg. The basic equations are those describing the steady state within each pool. Input equals output,

Pool *a*:

$$F_{ao} + F_{ab} = F_{ba} = F_{aa}$$

or

$$F_{ao} + k_{ab} Q_b = k_{ba} Q_a = k_{aa} Q_a$$

Pool *b*:

$$F_{ba} = F_{ab} + F_{ob} = F_{bb}$$

or

$$k_{ba} Q_a = k_{ab} Q_b + k_{ob} Q_b = k_{bb} Q_b$$

(Note that equalities also exist vertically for each respective pool.) Within these expressions are equalities to calculate Q_b and all rates. For example,

$$k_{ba} Q_a = k_{bb} Q_b$$
$$(0.03)(2.0) = 0.015 \, Q_b$$
$$Q_b = 4.0 \text{ mg}$$

Note that if Q_a is *not* known, the ratio Q_b/Q_a is predicted to be $k_{ba}/k_{bb} = 2$. Another equality to obtain F_{ab} is

$$F_{ab} = k_{ab} Q_b$$
$$= (0.005)(4.0) = 0.02 \text{ mg/min.}$$

The remaining calculable values are $F_{ao} = 0.04$, $F_{ba} = 0.06$, $F_{ob} = 0.04$.

12. Of great importance is the fact that the observed curve for pool *a* is compatible with every possible variant of an interchanging system illustrated in Figure 18. An example of solution will be given for the one with a side pool (Figure 18B). Equations (4) and (5) are the same for all models. Hence k_{aa} and k_{bb} are the same as previously calculated. But now

$$k_{ab} = k_{bb} = 0.015$$

Substitution and rearrangement of Eq. (6) gives

$$k_{ba} = \frac{(0.03)(0.015) - (0.037)(0.0081)}{0.015} = 0.01$$

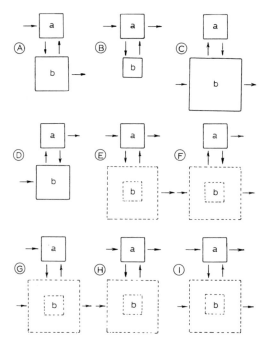

FIG. 18. Possible variants of an open two-pool exchanging system. In Models E–I the dashed boxes indicate that the size can vary over wide limits (see Table II).

From the model,

$$k_{oa} = k_{aa} - k_{ba} = 0.03 - 0.01 = 0.02$$

Flow rates:

$$F_{ba} = k_{ba} Q_a = (0.01)(2.0) = 0.02 \text{ mg/min}$$
$$F_{ab} = F_{ba} = 0.02 \text{ mg/min}$$
$$F_{oa} = k_{oa} Q_a = (0.02)(2.0) = 0.04 \text{ mg/min}$$

Because $F_{ab} = k_{ab} Q_b$,

$$Q_b = F_{ab}/k_{ab} = 0.02/0.015 = 1.33$$

Note the difference in internal rates and the size of pool b as compared to the prior model.

TABLE II

Values for Models of Figure 18

Fig. 18 model	$H_1, H_2, g_1,$ $g_2, k_{aa},$ k_{bb}	k_{oa}	k_{ob}	k_{ba}	k_{ab}	F_{ao}	F_{ba}	F_{oa}	F_{bo}	F_{ob}	F_{ab}	Q_b	Coefficients $\pm K$ for q_b/q_{ao}	Coefficients $\pm K/Q_b$ for SA_b
A	Identical	0	0.01	0.03	0.005	0.04	0.06	0	0	0.04	0.02	4.0	1.04	0.26
B	for	0.02	0	0.01	0.015	0.04	0.02	0.04	0	0	0.02	1.33	0.35	0.26
C	all	0	0.01	0.03	0.005	0	0.06	0	0.12	0.12	0.06	12.0	1.04	0.086
D	models	0.02	0	0.01	0.015	0	0.02	0.04	0.04	0	0.06	4.0	0.35	0.086
E	(see	a	a	a	a	0.04	a	a	0	a	0.02	a	a	0.26
F	paragraphs	b	b	b	b	0	b	b	b	b	0.06	b	b	0.086
G	5	0	0.01	0.03	0.005	c	0.06	0	c	c	c	c	1.04	c
H	and	0.02	0	0.01	0.015	d	0.02	0.04	d	0	d	d	0.35	d
I	10)	All values range between the above extremes												

[a] Values range between those for Models A and B.
[b] Values range between those for Models C and D.
[c] Values range between those for Models A and C.
[d] Values range between those for Models B and D.

PREDICTION OF TIME CURVE FOR POOL b

13. Solely from the foregoing information obtained from pool a the time curve for fraction of dose in pool b (Eq. (3b)) is predictable via Eq. (19) of Appendix II after the latter is simplified by making components relating to the nonexistent third pool equal to zero. The coefficients for Eq. (3c) are

$$K_1 = k_{ba}/(g_2 - g_1) \tag{7a}$$

$$K_2 = k_{ba}/(g_1 - g_2) \tag{7b}$$

As previously noted, K_1 and K_2 are of equal magnitude but of opposite algebraic sign. Slopes will be the same as those for the curve from pool a. Thus, for the model of Figure 14,

$$K_1 = \frac{0.03}{0.0081 - 0.037} = -1.04$$

With K_2 being $+1.04$, Eq. (3b) evolves. To convert to a curve for SA divide the coefficients K_1 and K_2 by pool size, 4, and Eq. (3a) is duplicated.

APPROACH TO OTHER MODELS

14. The solution of the remaining models in Figure 18 proceeds via algebraic methods similar to those demonstrated for Models A and B, however Models E, F, and I which have two routes of output, cannot be solved for explicit rate constants. The four unknown rate constants exceed the number of working equations. In Table II the allowable values comprising the families of possible solutions for E and F are seen to lie between those of Models A and B, and C and D, respectively (wherein one of the output rates, F_{oa} or F_{ob}, is zero). For Model I the upper and lower limits of possible rate constants are those for Models A and B, but possible *rates* may vary within the limits for A, B, C, and D. Models G and H, which have two sites of input but only one for output, can be solved for all rate constants but cannot be solved for Q_b, F_{ao}, and F_{bo} unless one of these three values is independently supplied. Any model of Figure 18 is amenable to explicit solution if, in addition to sampling pool a for SA, a similar sampling is also made on pool b, and the size of pool b (Q_b) is independently known. For demonstration the model will be assumed to possess all possible channels of input and output (Model I). An experimentally derived SA curve for species b subjected to graphic analysis, has a zero time intercept of 0.26 for the extrapolated tail. Units are fraction of dose per milligram carbon. The size of pool b is known to be 2.32 mg. To convert the intercept value to one for fraction of dose in the pool: $(0.26)(2.32) = 0.6$. By definition, this is the coefficient K_2 of Eq. (3c). Then, from Eq. (7b),

$$k_{ba} = (0.037 - 0.0081)(0.6) = 0.017$$

With k_{aa} and k_{bb} having been derived from the curve for pool a the calculation proceeds as follows:

$$k_{oa} = k_{aa} - k_{ba} = 0.03 - 0.017 = 0.013$$

From Eq. (6),

$$k_{ab} = 0.0087$$

and

$$k_{ob} = k_{bb} - k_{ab} = 0.015 - 0.0087 = 0.0063$$

For rates multiply rate constants by respective pool sizes,

$$F_{ba} = 0.035, \qquad F_{oa} = 0.025, \qquad F_{ab} = 0.02, \qquad F_{ob} = 0.015$$

Steady-state equations give

$$F_{ao} = (F_{ba} + F_{oa}) - F_{ab} = 0.04$$
$$F_{bo} = (F_{ab} + F_{ob}) - F_{ba} = 0$$

This is actually model E because $F_{bo} = 0$, however no constraint that any channel should be zero was initially imposed.

SIZE AND TURNOVER IN A UNIQUE MODEL

15. Steele (1964) has shown that if a two-pool model has both input and output to each pool (Model I of Figure 18) a certain set of values for the six rates will make possible the calculation of overall turnover rate $(F_{oa} + F_{ob})$ and overall size $(Q_a + Q_b)$ by a maneuver which, as a general approach, was warned against in Chapter 1. The tail of the SA curve when extrapolated to its zero time intercept (I_2) represents a simple exponential function $(I_2 e^{-g_2 t})$ which is the same as that of a *single* instantaneously mixed pool of size the same as $Q_a + Q_b$ and turnover rate the same as $F_{oa} + F_{ob}$. In other words, overall size is given by dividing dose by the SA at the zero time intercept of the tail,

$$Q_a + Q_b = q_{ao}/I_2 \tag{8a}$$

and turnover rate is given by multiplying overall size by slope, g_2 (rate constant for an assumed single pool),

$$F_{oa} + F_{ob} = (Q_a + Q_b)g_2 \tag{8b}$$

Values to satisfy such a unique set of rates and sizes which permit this simple maneuver may be predicted by pretending that the dose is indeed instantaneously mixed at t_0 thoughout both pools. Then, from Eq. (8a) applied to the SA curve of Figure 15, the joint size of a pretended single pool must be

$$Q_a + Q_b = 1/0.12 = 8.33 \text{ mg}$$

To satisfy Eq. (8b)

$$F_{oa} + F_{ob} = (8.33)(0.0081) = 0.068 \text{ mg/min}$$

It is known from the observed curve from pool a that Q_a must be 2 (paragraph 5). Thus $Q_b = 8.33 - 2 = 6.33$ mg. Now express total output rate in terms of the two pools,

$$k_{oa} Q_a + k_{ob} Q_b = 0.068 \text{ mg/min}$$

or

$$(k_{aa} - k_{ba})Q_a + (k_{bb} - k_{ab})Q_b = 0.068$$

The constants k_{aa} and k_{bb} are fixed by the curve (paragraph 10). Thus

$$(0.030 - k_{ba})2 + (0.015 - k_{ab})6.33 = 0.068 \tag{8c}$$

From Eq. (6),

$$(0.03)(0.015) - k_{ab} k_{ba} = 0.0003 \tag{8d}$$

Equations (8c) and (8d) are simultaneous equations which give

$$k_{ba} = 0.0219, \qquad k_{ab} = 0.00686$$

By difference k_{oa} and k_{ob} are both found to be 0.00814. Steady-state equations prove the following symmetrical configuration for the rates,

$$F_{ao} = F_{oa} = 0.0163$$
$$F_{bo} = F_{ob} = 0.0519$$
$$F_{ba} = F_{ab} = 0.0437$$

The equality of input and output for each pool, and of rate of interchange respectively, is characteristic for the set of values which apply to *any* observed SA curve from pool a in such a model. Further algebra establishes additional general relationships:

$$k_{ba} = H_1(g_1 - g_2), \qquad k_{ab} = H_2(g_1 - g_2)$$
$$k_{oa} = k_{ob} = g_2, \qquad F_{ao} = Q_a g_2, \qquad F_{bo} = Q_b g_2$$

This special case of Model I (demanding the particular equalities noted above) may be viewed as a curious exception to the rule that size and turnover rate of a two-pool system cannot be calculated by the extrapolated tail of an SA curve.

Solution by Sampling Secondary Pool *b* Alone

16. If experimental observations are limited to sampling the SA of pool *b* the curve must be normalized to provide coefficients (K_1 and K_2) for an equation in terms of fraction of dose present in the pool (paragraph 8). This ordinarily might be accomplished by knowing pool size (Q_b). Then Eq. (7) gives k_{ba} directly. If the model is such that $k_{aa} = k_{ba}$ this permits the same series of computations as before, i.e., when k_{aa} was obtained by sampling pool *a*. When k_{ba} and k_{aa} are not identical (e.g., Figure 18B) the quadratic equation relationship arising from Eqs. (5) and (6) will give a choice of two possible sets of values for the remaining rate constants. For Figure 18B one set will be the same as given by sampling pool *a* (Table II). For the other set, $k_{aa} = 0.025$, k_{bb} and $k_{ab} = 0.020$, $k_{oa} = 0.015$, and $Q_b = 1.0$. Of interest is the fact that this second set also is potential companion to an entirely different curve for pool *a* rather than the one at hand. Working backward the curve is shown to be

$$q_a/q_{a0} = 0.584e^{-0.037t} + 0.416e^{-0.0081t}$$

A Noninterchanging System

SINGLE ENTRANCE AND EXIT

Solution

17. Model A of Figure 19, having no reflux back to *a* from *b*, yields a simple exponential curve for SA of pool *a*. Its equation is $SA_a = 1.0e^{-0.02t}$. With dose called 1.0, the observed intercept identifies Q_a as 1 mg. The rate constant k_{ba} (or k_{aa}) is the slope, 0.02, and total inflow–outflow obviously is 0.02 mg/min. To complete the analysis pool *b* also must be sampled because without reverse flow pool *a* cannot "see" pool *b*. The SA curve of pool *b* is $SA_b = -1.0e^{-0.02t} + 1.0e^{-0.01t}$. One of the two slopes is contributed by pool *a* and the other by *b*. The former was shown to be 0.02. Being the larger, it is called g_1, leaving 0.01 for g_2. Substituting known values in Eq. (5) confirms that k_{bb} (or k_{ob}) $= g_2 = 0.01$. Then $Q_b = F_{ob}/k_{ob} = 2$ mg. The two intercepts of the SA curve (-1 and $+1$) multiplied by 2 will give the coefficients K_1 and K_2 for the normalized equation for fraction of dose in *b*.

Equivalence to "mother–daughter" equation

18. With K_1 and K_2 differing only in sign, Eqs. (7a) and (7b) may be substituted in Eq. (3c) to give an expression for activity in pool *b*,

$$q_b = q_{a0} \frac{k_{ba}}{g_2 - g_1} (e^{-g_1 t} - e^{-g_2 t}) \qquad (9)$$

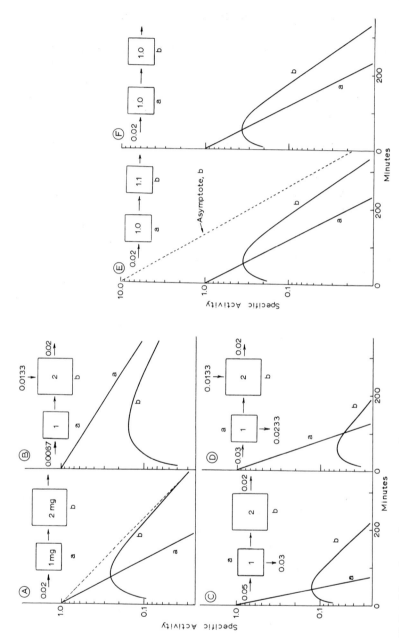

FIG. 19. Possible variants of an open, two-pool, one-way system. For conversion to curves for q_b versus time, multiply SA_b by Q_b. For rate constants of efflux divide rates as shown by donor pool size (in boxes). The dashed line in A is the asymptote approached by curve b. The equation for SA_b in Model E is $-10e^{-0.02t} + 10e^{-0.0182t}$. Other equations are given in the text.

Now, g_1 and g_2 are the same as k_{ba} and k_{ob}, but which slope goes with which rate constant is not known. The slope g_1 is, by definition, the larger. It will correspond to k_{ba} only if pool a is smaller than pool b. With flow through pools a and b being the same, and rate constants being F/Q, each rate constant is inversely proportional to pool size. With the uncertainty of correspondence in mind note that the expression derived by direct integration of a differential equation for rate constants in Model A is

$$q_b = q_{a0} \frac{k_{ba}}{k_{ob} - k_{ba}} (e^{-k_{ba}t} - e^{-k_{ob}t}) \tag{10}$$

Although k_{ba} appears to be the g_1 and k_{ob} the g_2 of Eq. (9), their identical position in the equations does not in itself prove numerical identity. It so happens that these two equations give the same numerical result because of the compensatory effect of the algebraic signs. Equation (10) is the classical equation for radioactive transformation where the decay of mother substance (a) produces daughter product (b). In terms of SA rather than quantity of tracer the common coefficient in Eq. (10) will embody k_{ob} in the numerator rather than k_{ba},

$$SA_b = SA_{o0} \frac{k_{ob}}{k_{ob} - k_{ba}} (e^{-k_{ba}t} - e^{-k_{ob}t}) \tag{11}$$

Equation (11) emerges by dividing Eq. (10) by Q_b, substituting F_{ba}/Q_a for k_{ba} in the numerator, then F_{ob} for F_{ba}. The common coefficient now contains q_{a0}/Q_a which is SA_{a0}, and F_{ob}/Q_b which is k_{ob}. By direct integration an equation companion to Eq. (10) is derivable for an unrestricted two-pool model having both input and output to each pool (Figure 19D). It is

$$q_b = q_{a0} \frac{k_{ba}}{k_{bb} - k_{aa}} (e^{-k_{aa}t} - e^{-k_{bb}t}) \tag{12}$$

ADDITION OF OTHER CHANNELS

Solutions

19. Equations for the SA curves of the next three models of Figure 19 are as follows:

Model B:

$$SA_a = 1.0e^{-0.0067t}$$
$$SA_b = -1.0e^{-0.01t} + 1.0e^{-0.0067t}$$

Model C:

$$SA_a = 1.0e^{-0.05t}$$
$$SA_b = -0.25e^{-0.05t} + 0.25e^{-0.01t}$$

Model D:

$$SA_a = 1.0e^{-0.03t}$$
$$SA_b = -0.167e^{-0.03t} + 0.167e^{-0.01t}$$

Model B theoretically is solvable, as was A, by sampling both pools. The listed SA equation for pool b predicts that the zero time intercept of the extrapolated tail will be 1.0. With k_{ba} determined to be 0.0067 via the curve from pool a, Eq. (7b) gives 2 for K_2. Then Q_b must be $K_2/1 = 2$. Values for F would follow as usual. In this particular example a practical problem would be encountered if an attempt were made to extrapolate the SA curve for pool b as drawn. With its two slopes being so nearly alike the actual pure *terminal* slope is long delayed in being reached. With the illustrated segment up to 400 min the zero intercept would come nowhere near 1.0. Models C and D are not solvable from the two SA curves alone. Solution of C is possible with foreknowledge of Q_b which with the intercept for the tail of the curve of SA_b, yields the coefficient K_2 and thence k_{ba}, and thence k_{oa} by difference from k_{aa} (which latter is the slope of the SA curve for pool a). Foreknowledge of F_{ob} or of F_{oa} likewise will permit complete solution. Model D is solvable from the two SA curves along with an independently known value for either Q_b or F_{ob}. Models C and D illustrate an interesting example of how the flow pattern in one pool affects the curve of another. The addition of exit F_{oa} makes k_{aa} different from k_{ba}. The effect in pool b is to change g_1 so that it becomes equivalent to k_{aa} rather than k_{ba}. Pool b "sees" $k_{ba} + k_{oa}$ rather than k_{ba} alone. Yet the coefficients, K, are influenced both by $k_{aa}(g_1)$ and k_{ba} (see Eq. (7)).

Ratio of F_{ba} to F_{bb}

20. Pool b in Models B and D receives labeled carbon from pool a (F_{ba}), and also nonlabeled, externally derived, carbon (F_{bo}). The ratio of F_{ba} to total input, i.e., $F_{ba}/(F_{ba} + F_{bo})$, or F_{ba}/F_{bb} is calculable from the SA curves of the two pools without resorting to any of the foregoing manipulations of compartment analysis. Note that in Models A and C, where pool b has no influx from the outside, the SA curve for pool b reaches its maximum value at the point where it crosses the curve of SA for pool a. But in Models B and D the peak of the curve for b (t_{max}) lies below the line for SA of pool a. The degree of depression of the peak below the point directly above it on the curve for pool a may be used to determine the relative contribution of nonlabeled carbon to pool b, i.e., to determine the ratio $F_{ba}/(F_{ba} + F_{bo})$. This ratio is simply the ratio of specific activities at the time of the peak, i.e. SA_b/SA_c at t_{max}. Proof of the relationship is presented in Appendix III. The ratio can be determined either by direct measure of SA_b and SA_a on the graph, or by calculating t_{max} after curve analysis by the formula as derived in Appendix III,

$$t_{max} = \frac{\ln(g_1/g_2)}{g_1 - g_2}$$

The derived time for t_{max} is substituted in the two equations for SA to compute SA_a and SA_b at this point in time.

TURNOVER RATE CONSTANTS EQUAL

21. All operations described to this point in the chapter are based on equations derived with the assumption that no two slopes are equal (Appendix II, A–C). If, in a nonreversible, two-pool system, the flow patterns, flow rates, and pool sizes are such that $k_{aa} = k_{bb}$, then $g_1 = g_2$. Thus, because of equal sized pools in Figure 19F, $k_{aa} = k_{bb} = 0.02$, which also is the common single slope, g. Although the equation for SA of pool a has the conventional form of $1.0e^{-0.02t}$, pool b takes a form which is strikingly different from any heretofore employed. With dose being 1,

$$SA_b = q_{a0}\, gte^{-gt} = (1)(0.02)(t)e^{-0.02t}$$

With $Q_b = 1$, the equation for q_b is the same. (See Appendix IID, Eq. (37).) This equation is peculiar in that it has no constant coefficient. As time increases the coefficient $0.02t$ increases. Its limit, i.e., the extrapolated intercept of the terminal downslope, is infinity. In effect, this means that the downslope cannot be extrapolated graphically. Also, a graphical estimate of slope is theoretically not accurate until infinite time has elapsed. Thus, the curve is not amenable to conventional compartment analysis based on curve peeling. But note that the differential equation for this model is exactly the same as for any other model,

$$dq_b/dt = -q_b k_{ob} + q_a k_{ba}$$

A computer, programmed to fit k_{ob} and k_{ba} to the observed curve, could work directly with this differential equation and thereby circumvent the theoretical impasse encountered with graphical analysis.

22. In Model E of Figure 19 the second pool has been made only slightly larger than the first. This means that g_2 is only slightly smaller than g_1. Note that the curve for SA_b is essentially indistinguishable from that for Model F, where the pools are of equal size. Also note that at 400 min the downslope is still far to the left of its asymptote ($10e^{-0.0182t}$). Much more time will be required before the terminal segment of the curve reaches this limiting line. If extrapolation is performed from the observed segment between 200 and 400 minutes the zero intercept will be 2 rather than 10 and the slope, g_2, will be 0.0165 rather than 0.0182. Peeling of the observed curve from the extrapolated line will give a value of 0.027 for g_1 rather than 0.02. The *apparent* equations for SA and quantity of tracer will be

$$SA_b = -2e^{-0.027t} + 2e^{-0.0165t}$$
$$q_b = -2.2e^{-0.027t} + 2.2e^{-0.0165t}$$

These are much different from the equations shown in the figure. But in spite of this discrepancy the errors tend to cancel so that Eqs. (7), when applied to this derived expression for q_b, give 0.023 for k_{ba}. The error is less than might be expected.

23.   If the observed curves are for quantity of tracer rather than for SA their relationship at a point of crossover will provide a clue that the two pools are the same size or nearly so. If the curves were for SA, the one from pool a should cross that for pool b at the peak of the latter curve (t_{max}) *regardless of relative size*. This is the point where $SA_a = SA_b$. Thereafter SA_b can only decline. However, for curves for *quantity* of tracer the crossover is not at t_{max} of pool b *unless the two pools are the same size*. This can be appreciated by starting with SA curves which cross at t_{max} then converting to quantity of tracer in each by multiplying each curve by the corresponding pool size. If Q_a does not equal Q_b the curves will move apart vertically and crossover will be shifted laterally.

Precautions and Limitations

THE NUMBER OF COMPARTMENTS

24.   Injudicious lumping of an actual three-compartment system into a working model with only two compartments introduces errors similar to those arising when two are injudiciously lumped into one (Chapter 1). If, in a supposed two-pool system having a supposed one-way flow (Figure 19), the semilogarithmic plot of SA versus time of pool a gives a prolonged curved line, this means either that the system is reversible or that another compartment exists and interchanges with pool a. If such a third pool can be reconciled with a known physiologic or chemical model the analysis could be for a three-compartment system as described in Chapter 3.

EXTRAPOLATION TO ZERO TIME

25.   In paragraph 5 a method was described whereby Q_a could be estimated by summing the intercept values of the "peeled" straight-line components

of a complex curve of SA. Although this is quite straightforward in hypo-
thetically pure models used for illustration, it may be difficult in practice.
One problem is to obtain data points sufficiently early and in good orderly
alignment to permit construction of the initial segment of the curve. Such
difficulty is compounded by the likelihood of encountering a rather steep
initial downslope. But even if an apparently orderly plot approaches close to
the zero time ordinate an even earlier and steeper (i.e., much more rapid)
component of a more complex curve may not be recognized. The question
being approached is: In a real experimental setting can the *size of the very
first* pool be estimated by extrapolation? The chances may be poor in
certain circumstances. For example, when ^{14}C glucose is injected intravenously into a rat and sampling is begun at 5 min, the complex plasma curve
beyond 5 min, when peeled and extrapolated, gives a pool size equal to the
entire pool for extracellular glucose (Shipley *et al.*, 1967). Yet, the real *primary*
pool is plasma. In this small animal the interchange between plasma, red
cells, and interstitial fluid is so rapid that this initial phase of tracer transport is essentially completed within 5 min. Earlier sampling might be
attempted hoping to resolve these early events but the chance of success is
remote. Diffusion is progressing rapidly even before mechanical mixing in
blood is completed. Mechanical mixing would have to be more rapid than is
possible to study the aforementioned rates of diffusion. In the present example
the glucose in a conglomerate of anatomic pools behaves as a single pool
as it participates in relatively slower processes which dominate the curve
after 5 min. The model of Figure 14 can be modified to illustrate this point.
Assume that pool *a* is actually made up of two subunits of approximately 1 mg
each and that flow rate from the very first (labeled) to the secondary sub-
unit is at least 0.2 mg/min. The curve of Figure 15 then would begin at 1.0
rather than 0.5, however its contour after 10 min would be substantially the
same as drawn. Inspection of the curve as shown will confirm the assertion
that the unobserved segment between 0 and 10 min can just as well be drawn
to 1.0 on the ordinate at $t = 0$ as it can be drawn to 0.5. This problem will
be discussed again in Chapter 4.

26. The size of the very first pool and the value of its rate constants, turnover, etc., may be difficult to estimate, not only because of rapid physical
events, but also because of rapid chemical transformation of the primary
material. For example, compound A, known to exist in plasma, may be considered to represent a primary pool. Assume that compound A is *rapidly*
and *reversibly* converted to compound B, also present in blood. After introducing tracer consisting of carbon-labeled compound A and following the
SA of carbon in A for long-term kinetic events, the apparent primary pool
will be the carbon in compounds A + B, not that in A alone.

NONUNIFORM LABELING

27.   Paragraph 2 emphasized the importance of designating SA and quantity of tracer in terms of tracer and tracee atoms rather than molecules. In a complex chemical system such as that of Model I of Figure 18 another stricture exists. Radioactive atoms in the introduced molecule must *not* be confined to a specific site (e.g., carbon 1 of glucose) if conversions, exits, or entrances (arrows on the diagram) represent passage of a fragment of the molecule possessing the site of label, but such fragment is not always accompanied by the same proportion of the other fragment. That is to say, if the fragment containing the site of label is partitioned unequally among the various routes of transfer, the only meaningful analysis would be in terms of pool content and rates for the labeled fragment alone. The other portion of the molecule would, in effect, be unrecognized as a part of the system being analyzed. If the tracer were ^{14}C the calculated rates, rate constants, and pool sizes would not apply to total carbon. The problem is obviated by using a tracer molecule which is uniformly and equally labeled at all six carbon sites. Fragmentation then will yield equal proportions of tracer and tracee atoms everywhere in the system.

CHAPTER 3

COMPARTMENT ANALYSIS: THREE-POOL OPEN SYSTEMS

A Completely Interchanging System

1. A three-pool system possessing all possible routes of interchange, ingress, and egress for each pool is shown in Figure 20A. This unrestricted version is the basis for the general equations of Appendix II. For explicit mathematical solution based on a time curve for tracer obtained from one of the pools, considerable restriction in number of flow channels is required (Models B–K). Calculations for such models are simplified when some of the terms of the general equations become zero and drop out. In the examples which follow the same curve for SA will be adopted for labeled pool *a* in Models B–I. Also, for simplicity, the size of pool *a* (Q_a) will be called one unit of mass in all models. Explicit calculations will be based on the curve for pool *a* because the labeled pool is most likely to be available for sampling. As was emphasized in Chapter 2, paragraph 2, although material in a chemical pool is of a particular molecular species, sizes and rates are in terms of tracee *atoms*. See also Chapter 2, paragraph 27, in connection with nonuniform labeling.

The Curve for Pool *a*

2. In all models of Figure 20, save J and K, a curve of SA or quantity of tracer in pool *a* will have three exponential components (as will curves for pools *b* and *c*). This is true for any restricted version of Model A provided

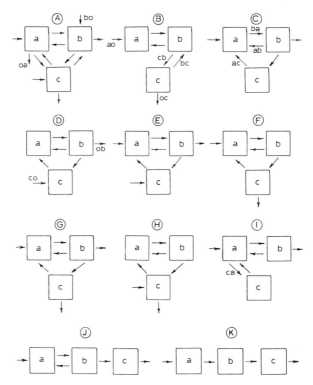

Fɪɢ. 20. An assortment of three-pool models used for compartment analysis. Letters accompanying representative transfer rates would be appropriate subscripts of either F (rate) or k (rate constant).

that no pool is eliminated and that interchange, whether direct or indirect, is retained between all pools. To avoid the awkwardness of small fractions which may be encountered in an experimental setting, SA conveniently may be expressed as percent of dose, rather than fraction of dose, per unit weight of tracee. The solid curve of Figure 21A, which is observed SA_a for models A–I, has ordinate values expressed as percent of dose. At zero time 100% of dose accompanies one unit of mass, and Q_a having been assigned a value of 1, the SA at zero time is also 100%. SA_a and q_a happen to be the same because of this assigned value (to simplify subsequent arithmetic). The curve of Figure 21A is drawn as though sampling began at 2 min. Deriving its equation by graphic analysis as described previously for a two-pool model in Chapter 2, the extrapolation of the terminal segment (dashed line) gives $10e^{-0.01t}$, and subtraction of values on this line from the original curve gives a two-component curve (solid line of Figure 21B). The extrapolated dashed line now is $20e^{-0.1t}$, and the final subtracted component (dot–dash) is $70e^{-0.5t}$. As a

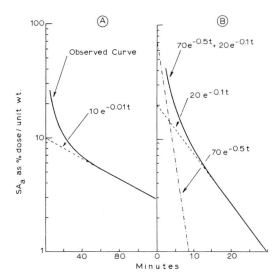

FIG. 21. Successive peeling of SA curve for pool *a*, models of Figure 20. Observed curve is solid line in A. See text for its equation. Dashed line is extrapolation of tail (third term). Part B shows emergence of first and second terms by a second extrapolation and subtraction.

practical point in passing, note how vulnerable the final subtraction would be to errors in defining an original experimental curve. A difference is being obtained between two relatively large values in the restricted interval between 2 and 10 min. Returning to the result of analysis (which in this case is exact because the curve was deliberately drawn from the equation) the curve of Figure 21A is

$$SA_a = 70e^{-0.5t} + 20e^{-0.1t} + 10e^{-0.01t} \qquad (1a)$$

As explained in Chapter 2, paragraph 6, this may be modified by expressing both sides as a ratio to the starting value at zero time

$$SA_a/100 = 0.7e^{-0.5t} + 0.2e^{-0.1t} + 0.1e^{-0.01t} \qquad (1b)$$

This now is in a normalized form wherein coefficients and exponential slopes have symbols which will appear in the compartment analysis to follow

$$SA_a/SA_{a0} = H_1 e^{-g_1 t} + H_2 e^{-g_2 t} + H_3 e^{-g_3 t} \qquad (1c)$$

CALCULATIONS FOR MODEL B OF FIGURE 20

Rate Constants

3. Equations for calculating rate constants are five in number. Equations (2) and (3), which follow, apply to all models in the series. Equation (4) is a rearrangement of Eq. (16) of Appendix II. Equations (5) and (6) have been

modified from general Eqs. (14) and (15) of Appendix II by eliminating terms containing k_{ca} and k_{ac} which are zero in this model,

$$H_1 g_1 + H_2 g_2 + H_3 g_3 = k_{aa} \tag{2}$$

$$g_1 + g_2 + g_3 = k_{aa} + k_{bb} + k_{cc} \tag{3}$$

$$k_{bb} k_{cc} - k_{bc} k_{cb} = H_1(g_2 - g_1)(g_3 - g_1) - g^2_1 + g_1(k_{bb} + k_{cc}) \tag{4}$$

$$g_1 g_2 + g_2 g_3 + g_3 g_1 = k_{aa}(k_{bb} + k_{cc}) + (k_{bb} k_{cc} - k_{bc} k_{cb}) - k_{ab} k_{ba} \tag{5}$$

$$g_1 g_2 g_3 = k_{aa}(k_{bb} k_{cc} - k_{bc} k_{cb}) - k_{cc} k_{ab} k_{ba} \tag{6}$$

The algebra is as follows,

$k_{aa} = 0.3710$	(from Eq. (2))
$k_{bb} + k_{cc} = 0.2390$	(substituting 0.371 in Eq. (3))
$k_{ba} = k_{aa} = 0.3710$	(fixed by the model)
$k_{bb} k_{cc} - k_{bc} k_{cb} = 0.00670$	(substituting in Eq. (4))
$k_{ab} k_{ba} = 0.0394$	(substituting foregoing calculated value in Eq. (5))
$k_{ab} = 0.1061$	(from foregoing product and derived value of k_{ba})
$k_{cc} = 0.05044$	(substituting in Eq. (6))
$k_{bb} = 0.1886$	(substitute k_{aa} and k_{cc} in Eq. (3))
$k_{bc} k_{cb} = 0.002813$	(from Eq. (4) and other derived values)
$k_{cb} = 0.0824$	($k_{cb} = k_{bb} - k_{ab}$)
$k_{bc} = 0.03410$	(from the previous two equalities)
$k_{oc} = 0.01634$	($k_{oc} = k_{cc} - k_{bc}$)

To achieve arithmetic balance the values are shown to more significant figures than would be anticipated in a biologic experiment.

Flow-rates and pool sizes

4.

$$F_{ba} = k_{ba} Q_a = 0.371$$
$$F_{bb} = F_{ba} + F_{bc} = 0.371 + F_{bc}$$

or

$$k_{bb} Q_b = 0.371 + k_{bc} Q_c$$

which is

$$0.1886 Q_b = 0.371 + 0.0341 Q_c$$

and

$$F_{cc} = F_{cb} \quad \text{or} \quad k_{cc} Q_c = k_{cb} Q_b$$

which is

$$0.05044Q_c = 0.0824Q_b$$

$$\begin{aligned}Q_b &= 2.790 \\ Q_c &= 4.565\end{aligned}$$ from the foregoing simultaneous equations

$$\begin{aligned}F_{cb} &= 0.230 \\ F_{ab} &= 0.296 \\ F_{oc} &= 0.0746 \\ F_{bc} &= 0.156\end{aligned}$$ from known rate constants and pool sizes

A check of these calculations should confirm that inflow balances outflow for each pool. For example: $F_{ba} + F_{bc} = F_{ab} + F_{cb}$. Moreover, the rate constants of efflux from each pool should add up properly, e.g., $k_{bb} = k_{ab} + k_{cb}$, etc.

The equation for activity in pool b

5. The equation for the amount of tracer (q_b) in pool b as a function of time is determined from Eqs. (19), (21)–(23) of Appendix II with the term $k_{bc}k_{ca}$ being zero,

$$K_1 = \frac{(-g_1 + k_{cc})k_{ba}}{(-g_1 + g_2)(-g_1 + g_3)} \tag{7}$$

$$K_2 = \frac{(-g_2 + k_{cc})k_{ba}}{(-g_2 + g_3)(-g_2 + g_1)} \tag{8}$$

$$K_3 = \frac{(-g_3 + k_{cc})k_{ba}}{(-g_3 + g_1)(-g_3 + g_2)} \tag{9}$$

The solutions are $K_1 = -0.851$, $K_2 = +0.511$, $K_3 = +0.340$. The three values of K must always add up to zero since at t_0, $q_b = 0$. Putting the coefficients into an equation for fraction of dose in pool b:

$$q_b/q_{a0} = K_1 e^{-g_1 t} + K_2 e^{-g_2 t} + K_3 e^{-g_3 t}$$

The slopes will be the same as in pool a. Bring q_{a0} to the right as a common coefficient,

$$q_b = 100(-0.851e^{-0.5t} + 0.511e^{-0.1t} + 0.340e^{-0.01t})$$

In converting this equation to one for specific activity (SA_b) the common coefficient is $100/2.79$ instead of 100.

The equation for activity in pool c

6. Coefficients (L) for pool c emerge from Eqs. (20)–(23) of Appendix II. The denominators for the three equations giving L_1, L_2, and L_3 are, respectively, the same as for the series of K coefficients in Eqs. (7)–(9). A common numerator applies to all three of the equations for L. Because k_{ca} is zero, the numerator of Eq. (20) of the appendix is simply $k_{ba}k_{cb}$. The derived L coefficients appear in the equation as follows,

$$q_c = 100(+0.156e^{-0.5t} - 0.850e^{-0.1t} + 0.694\,e^{-0.01t})$$

In the corresponding equation for SA_c the common coefficient is $100/4.57$ instead of 100. Note that the L coefficients also add up to zero. Curves of q and SA for all three pools of Model B are shown in Figures 22A and B. The identical contour but shift in vertical position upon conversion from q to SA for pool b and c is apparent at a glance.

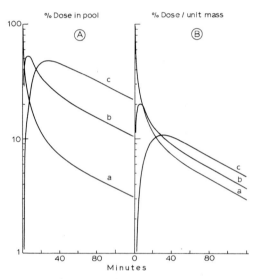

FIG. 22. (A) Curves for quantity of tracer as percentage of dose. (B) Curves for SA as percentage of dose per unit mass. The model is Figure 20B.

Overall check

7.   A check on the correctness of the derived values for the coefficients K and L, and for total outflow from the system, may be performed by a method based on a principle to be described in Chapter 5. Total outflow (F_{oc}) is equal to the dose divided by the integral from zero to infinity of the specific activity curve for any compartment of a system into which all input enters through the labeled pool. For pool b the integral to ∞ is

$$\frac{100}{2.79}\left(\frac{K_1}{g_1} + \frac{K_2}{g_2} + \frac{K_3}{g_3}\right)$$

which is 1340. (See paragraph 5 for explicit values.) For pool a or pool c the same value (1340) is obtained if 100 and coefficients H (paragraph 2) or 100/4.57 and coefficients L (paragraph 6) are used, respectively. Then

$$F_{oc} = \frac{\text{dose}}{1340} = \frac{100}{1340} = 0.0746$$

SOLUTION FOR OTHER MODELS IN FIGURE 20

Model C

8.

$$k_{aa} = 0.371$$
$$k_{bb} + k_{cc} = 0.239 \quad \text{(as with Model B)}$$
$$k_{ba} = 0.371 \quad \text{(from the model)}$$

Equations (16), (14), and (15) of Appendix II, modified by deletion of rate constants of nonexistent channels will give the following,

$$k_{bb}k_{cc} = H_1(g_2 - g_1)(g_3 - g_1) - g_1{}^2 + g_1(k_{bb} + k_{cc}) = 0.00670 \quad (10)$$

$$g_1g_2 + g_2g_3 + g_3g_1 = k_{aa}(k_{bb} + k_{cc}) + k_{bb}k_{cc} - k_{ab}k_{ba} = 0.0560 \quad (11)$$

$$g_1g_2g_3 = k_{aa}k_{bb}k_{cc} - (k_{cc}k_{ab}k_{ba} + k_{ac}k_{cb}k_{ba}) = 0.0005 \quad (12)$$

$$k_{bb}^2 - 0.239k_{bb} + 0.0067 = 0$$

or

$$k_{cc}^2 - 0.239k_{cc} + 0.0067 = 0$$

(from Eq. (3) for $k_{bb} + k_{cc}$ and Eq. (10) for $k_{bb}k_{cc}$).
   The quadratic solution gives

$$k_{bb} \text{ or } k_{cc} = 0.20655 \quad \text{or} \quad 0.03245$$
$$k_{ab} = 0.1061 \quad \text{(from Eq. (11))}$$
$$k_{ac}k_{cb} = 0.001909 \quad \text{(from Eq. (12) when } k_{cc} = 0.03245)$$

If $k_{cc} = 0.2065$, substitution in Eq. (12) produces a negative (not allowed) value for the product $k_{ac}k_{cb}$. Thus

$$k_{cc} = 0.03245$$
$$k_{bb} = 0.20655$$
$$k_{ac} = 0.03245 \quad \text{(identical to } k_{cc} \text{ in this model)}$$
$$k_{cb} = 0.05883 \quad \text{(product } k_{ac}k_{cb} \text{ previously determined)}$$
$$k_{ob} = 0.04160 \quad \text{(from } k_{bb} - (k_{ab} + k_{cb}))$$
$$Q_b = 1.796 \quad \text{(because } Q_b k_{bb} = Q_a k_{aa})$$
$$F_{ob} = 0.0746, \quad F_{ao} = 0.0746, \quad F_{ab} = 0.1906, \quad F_{cb} = 0.1057$$
$$Q_c = 3.25 \quad \text{(from } Q_c = F_{cc}/k_{cc} = F_{cb}/k_{cc})$$
$$F_{ac} = 0.1057 \quad \text{(from } k_{ac} Q_c)$$

Equations to determine the coefficients K or L are the same as those used for Model B. Derived values are shown in Table III and the SA curves in Figure 23C.

Model D

9. This model differs from Model C in only one respect. Inflow is via pool c rather than pool a. Note that a change in location of inflow has no effect on any of the existing rate constants for interflow or outflow because the calculations of paragraphs 8 are identical except for the size of pool c and therefore F_{ac}.

$$F_{co} = F_{ob} = 0.0746$$
$$F_{cb} = Q_b k_{cb} = 0.1057$$
$$F_{cc} = F_{co} + F_{cb} = 0.1803 = F_{ac}$$
$$Q_c = F_{cc}/k_{cc} = 5.55$$

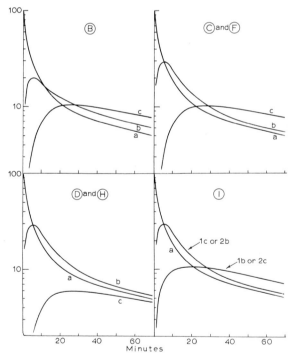

FIG. 23. Comparison of SA curves (percentage of dose per unit mass) for the indicated models of Figure 20. In Model I the curve for pool b, Sol. 1, is the same as that for pool c, Sol. 2, and vice versa.

TABLE III

VALUES FOR MODELS OF FIGURE 20

Model	B	C	D	E	F	G	H	I-1	I-2
Q_b	2.79	1.80	1.80	1.80	1.80	1.80	1.80	6.17	1.19
Q_c	4.57	3.25	5.55	3.25–5.55	5.55	3.25–5.55	9.50	0.827	3.86
k_{ba}	0.371	0.371	0.371	0.371	0.371	0.371	0.371	0.200	0.246
k_{ab}	0.106	0.106	0.106	0.106	0.106	0.106	0.106	0.020	0.144
k_{cb}	0.082	0.059	0.059	0.059	0.100	0.059–0.100	0.100	—	—
k_{bc}	0.034	—	—	—	—	—	—	—	—
k_{ca}	—	0.032	0.032	0.032	0.019	0.019–0.032	0.019	0.171	0.125
k_{ac}	—	0.042	0.042	0.042	—	0–0.042	—	0.207	0.032
k_{ob}	—							0.012	0.063
k_{oc}	0.016	—	—	—	0.013	0–0.013	0.013	—	—
F_{ba}	0.371	0.371	0.371	0.371	0.371	0.371	0.371	0.200	0.246
F_{ab}	0.296	0.191	0.191	0.191	0.191	0.191	0.191	0.123	0.171
F_{cb}	0.230	0.106	0.106	0.106	0.180	0.106–0.181	0.180	—	—
F_{bc}	0.156	—	—	—	—	—	—	—	—
F_{ca}	—	0.106	0.180	0.106–0.180	0.106	0.106	0.180	0.171	0.125
F_{ac}	—	0.075	0.075	0.075	—	0–0.075	—	0.171	0.125
F_{ob}	—					0–0.075		0.075	0.075
F_{oc}	0.075	—	—	—	0.075	−0.075	0.127	—	—
K_1	−0.851	−0.885	−0.885	−0.885	−0.885	−0.885	−0.885	−0.300	−0.586
K_2	+0.511	+0.696	+0.696	+0.696	+0.696	+0.696	+0.696	−0.592	+0.461
K_3	+0.340	+0.189	+0.189	+0.189	+0.189	+0.189	+0.189	+0.892	+0.125
L_1	+0.156	+0.111	+0.111	+0.111	+0.190	+0.111–+0.190	+0.190	−0.408	−0.188
L_2	−0.850	−0.604	−0.604	−0.604	−1.035	−0.604–1.305	−1.035	+0.321	−0.371
L_3	+0.694	+0.493	+0.493	+0.493	+0.845	+0.493–+0.845	+0.845	+0.087	+0.559

Although the curves for q_b and q_c are unchanged from Model C the specific activity of pool c will at all times be lower by the ratio 3.25/5.55. (See Figure 23D.)

Model E

10. In this model the inflow is shared by both pool a and pool c instead of being confined to one or the other as in Models C and D. Although rate constants are calculable as in C and D, an explicit solution for F_{ao}, F_{co}, F_{ac}, and Q_c is not possible unless one of the four is known, or the ratio of F_{ao} to F_{co} is known. The limits of the range of possible values for Q_c are 3.25 when F_{co} is zero (Model C) and 5.55 when F_{ao} is zero (Model D).

Model F

11. The change from Model C is that effluent is from pool c rather than pool b. Calculations differ from those for Model C as follows,

$$k_{cb} = k_{bb} - k_{ab} = 0.1004$$
$$F_{cc} = F_{cb} = Q_b k_{cb} = 0.1804$$
$$Q_c = F_{cc}/k_{cc} = 5.55$$
$$k_{ac} k_{cb} = 0.001909 \qquad \text{(from Eq. (12))}$$
$$k_{ac} = 0.01901$$
$$k_{oc} = k_{cc} - k_{ac} = 0.0134$$
$$F_{oc} = 0.0746 = F_{ao}$$

Model G

12. Whereas all rate constants were calculable when two sources of *input* were present (Model E) such is no longer the case where two routes of *output* are present. The limits, as listed in Table III, are the values for Models C and F where output is confined to either F_{ob} or F_{oc}.

Model H

13. The pattern now is similar to that of Model F except that both inflow and outflow are connected to pool c,

$$F_{ac} = F_{cb} = 0.1804$$
$$Q_c = F_{ac}/k_{ac} = 9.50$$
$$F_{oc} = Q_c k_{oc} = 0.1273 = F_{co}$$

Model I

14. This model illustrates a circumstance in which, although explicit solution is possible, a step in calculation involving a quadratic equation yields two equally compatible sets of explicit values which apply to a variety of rate constants and to the size of pools b and c. The algebra yields values identical to those of Model C until after the solution of the quadratic. Thus

$$k_{bb} = 0.03245, \qquad k_{cc} = 0.20655 \qquad \text{(Solution 1)}$$
$$k_{bb} = 0.2065, \qquad k_{cc} = 0.03245 \qquad \text{(Solution 2)}$$

Equation (14) of Appendix II becomes

$$0.0560 = k_{aa}k_{bb} + k_{aa}k_{cc} + k_{bb}k_{cc} - k_{ab}k_{ba} - k_{ac}k_{ca} \tag{13}$$

Equation (15) of Appendix II becomes

$$0.0005 = k_{aa}k_{bb}k_{cc} - k_{bb}k_{ac}k_{ca} - k_{cc}k_{ab}k_{ba} \tag{14}$$

By substituting the values for k_{bb} and k_{cc} as listed for Solution 1, and using k_{cc} for k_{ac}, Eqs. (13) and (14) each may be set equal to $k_{ab}k_{ba}$ and thereby to each other, leaving only k_{ca} as unknown,

$$k_{ca} = 0.1709$$
$$k_{ba} = k_{aa} - k_{ca} = 0.2001$$

From Eq. (13) or (14),

$$k_{ab}k_{ba} = 0.004070$$
$$k_{ab} = 0.02034$$
$$Q_b = F_{bb}/k_{bb} = F_{ba}/k_{bb} = 6.17$$
$$Q_c = F_{cc}/k_{cc} = F_{ca}/k_{cc} = 0.827$$
$$k_{ob} = k_{bb} - k_{ab} = 0.012$$

Corresponding values for Solution 2 are given in Table III. In practice the choice between these two solutions would depend on the compatibility of Q_b, Q_c, and/or the several rates of interchange with independently derived information.

SAMPLING POOL b OR POOL c, ANY MODEL

15. When pool b or pool c is the exclusive site of sampling, and its *size is independently known*, an observed SA curve will yield values for the intercept-coefficients, K or L, to be employed in appropriate equations of Appendix II. If the model embodies k_{ba}, Eqs. (13)–(15) and (19), (21)–(23) of Appendix II will lead to a complete solution via the curve from pool b (coefficients, K). If k_{ca} exists and the observed curve is from pool c (coefficients, L) the explicit solution is reached via Eqs. (13)–(15) and (20)–(23) of Appendix II. Absence of k_{ba} in equations for K, or of k_{ca} in equations for L, will mean that no two equations are independent, and an explicit solution, therefore, is not possible.

TABLE IV

SOLUTIONS GIVEN BY SAMPLING VARIOUS POOLS, MODEL I, FIGURE 20

	(1)	(2)	(3)	(4)	(5)	(6)	(7)	(8)	(9)	(10)	(11)	(12)
	Sample pool a		Sample pool b, find K values of Sol. 1, given by pool a		Sample pool c, find L values of Sol. 1, given by pool a			Sample pool b, find K values of Sol. 2, given by pool a		Sample pool c, find L values of Sol. 2, given by pool a		
	Sol. 1	Sol. 2	1	2	1	2	3	1	2	1	2	3
H_1	0.7000	0.7000				0.6641			0.8160		0.6929	
H_2	0.2000	0.2000				0.2283			0.1087		0.1860	
H_3	0.1000	0.1000				0.1077			0.0753		0.1210	
K_1	−0.2996	−0.5857				−0.2628			−0.5857		−0.5686	
K_2	−0.5924	+0.4608	Same as "sample pool a, 1"	Solution excluded by a negative derived value for a rate constant	Same as "sample pool a, 1"	−0.6252	Solution excluded by a negative derived value for a rate constant	Same as "sample pool a, 2"	+0.4608	Same as "sample pool a, 2"	+0.4204	Solution excluded by a negative derived value for a rate constant
K_3	+0.8920	+0.1249				+0.8880			+0.1249		+0.1483	
L_1	−0.4077	−0.1879				−0.4077			−0.3039		−0.1879	
L_2	+0.3207	−0.3715				+0.3207			−0.2802		−0.3715	
L_3	+0.0869	+0.5593				+0.0869			+0.5841		+0.5593	
Q_a	1.0000	1.0000				1.0729			0.7645		1.1454	
Q_b	6.1697	1.1886				6.1213			1.1886		1.3352	
Q_c	0.8273	3.8689				0.8273			4.1044		3.8689	
k_{aa}	0.3710	0.3710				0.3559			0.4196		0.3663	
k_{bb}	0.0324	0.2066				0.0324			0.1579		0.2066	
k_{cc}	0.2066	0.0324				0.2216			0.0324		0.0372	
k_{ba}	0.2001	0.2455				0.1851			0.2455		0.2408	
k_{ab}	0.0203	0.1438				0.0202			0.0951		0.1507	
k_{ca}	0.1709	0.1255				0.1709			0.1741		0.1255	
k_{ac}	0.2066	0.0324				0.2216			0.0324		0.0372	
k_{ob}	0.0121	0.0628				0.0122			0.0628		0.0559	
F_{ba}	0.2001	0.2455				0.1986			0.1877		0.2758	
F_{ab}	0.1255	0.1709				0.1239			0.1131		0.2012	
F_{ca}	0.1709	0.1255				0.1834			0.1331		0.1437	
F_{ac}	0.1709	0.1255				0.1834			0.1331		0.1437	
F_{ob}	0.0746	0.0746				0.0746			0.0746		0.0746	

Thus, save for Model G (which has an excessive number of unknown rate constants), Models B–H of Figure 20 qualify for a single explicit solution by sampling SA in pool b of known size. However these same models cannot be solved by sampling pool c alone because k_{ca} is absent. Model I is a special case. As was true for pool a, pool b can give more than one acceptable solution. If the model is such that it is compatible with Solution 1 obtained via pool a (column 1 of Table IV) the K coefficients of column 1 will emerge. Via these coefficients, a quadratic step then will yield values identical to those of column 1 (column 3), or an incompatible result (column 4). If the model is such that it is compatible with solution 2 obtained via pool a (column 2) these K coefficients will yield two possible solutions (columns 8 and 9). The latter one is completely new. If pool c is sampled a cubic equation gives the results shown in columns 5–7 or 10–12 depending on whether the mated values are in column 1 or 2. Now, another completely new set of possible values emerges (column 11). One purpose in presenting the table is to illustrate the abundance of alternate solutions. Even so, the list is not complete. For example, the new set of H coefficients for pool a in columns 6, 9, and 11 each may open up new possibilities.

16. If a curve from one pool does not give an explicit solution, knowledge of the size of an additional pool may suffice to provide such a solution. Sampling an additional pool is generally a more powerful device. Model 20G cannot be solved for all rate constants via the curve from pool a alone but this is possible when an additional equation for quantity of tracer in pool b or pool c provides coefficients K or L to use in supplementary equations. Also, in Model I where the curve from one pool gives a choice of two possible solutions the pertinent one may be identified by finding the matching set of values given via another pool (Table IV). Another approach for achieving complete solution of certain models is to label an alternate pool on a separate occasion. This is especially advantageous because solution via a *primary* labeled pool tends to be more explicit than via a secondary pool. In certain chemical systems, e.g., steroid compounds, it may be possible to label two species simultaneously with separate tracers, e.g., ^{14}C for one, and ^{3}H for the other. But this is possible only when the two are not partitioned apart in chemical transformation, that is to say both are stably bound to the molecule and kinetic events may be viewed in terms of the main molecular nucleus.

Three Compartments Not Interchanging Completely

MODEL J (FIGURE 20)

17. Because pool c is within a one-way path it has no effect on the curves observed in pools a and b. Analysis for these first two pools as an interchanging complex is described in Chapter 2. As was the case for Model B, the entire three-pool system cannot be solved by sampling pool c even though one less rate constant exists. Curve SA_c will have three exponential components. If Q_c is known this will yield the L coefficients from the observed intercepts. Available equations are four in number: the equivalent of only one for L, plus Eqs. (3), (5), and (6) of paragraph 3. Unknowns are four in number: k_{ba}, k_{ab}, k_{cb}, and k_{oc}. Solution might appear to be possible, but it is not. The equations for L and what remains of Eqs. (5) and (6) are identical expressions in terms of the product $k_{ba}k_{cb}$. Solution is not possible with only two independent equations (Eq. (3), and either L, Eq. (5), or Eq. (6)).

MODEL K (FIGURE 20)

18. With injection of tracer into pool a the SA curve for pool a is a " straight-line " single exponential, that for pool b is a double exponential, and that for pool c, triple. In this one-way system the three slopes (g_1, g_2, g_3) observed in pool c correspond to the three rate constants (k_{aa}, k_{bb}, k_{cc}) but not necessarily in this order. For example, g_1 does not necessarily represent k_{aa}. If pool c should be the smallest of the three then $g_1($ the steepest slope by definition) will represent k_{cc} because $F_{aa} = F_{bb} = F_{cc}$, and $k_{cc} = F_{cc}/Q_c$. If Q_c and rate of flow through the system are known this will give k_{cc} and identify the accompanying slope which is numerically the same. By substituting this value for the appropriate g (and k_{cc}) in Eqs. (3), and (5) or (6), a quadratic step will yield an "either or" for k_{aa} (and Q_a) versus k_{bb} (and Q_b). Hence, pool c "sees" two preceding pools differing in size and rate constant, but it cannot tell their sequence. If all three pools are the same size, the system requires a special analysis as outlined in Chapter 2, paragraph 21 (and Chapter 7, paragraph 8).

Dissection of Rates

TURNOVER

19. The *turnover rate* for a given pool within a complex system encompasses total rate of input (balanced by total output). Whether conceived as rate of input or output it is the total of *all* respective rates including reversible communications with other pools. For pool a it is F_{aa}, which in Figure 24A is the same as F_{ba}, or $F_{ab} + F_{ac} + F_{ao}$, i.e., 55 mass units of pool a atoms per unit time. For pool b it also is 55, while for pool c it is 80. This rate must be distinguished from net production rate to be described later. Turnover rate includes that for recycled atoms not new to the pool. As was the case for a single pool, turnover time is the time required for the movement through the pool of an amount of tracee equal to pool size. For pool a this is Q_a/F_{aa}. If Q_a is assigned a value of 100 mg and time is in minutes, turnover time $=$ $100/55 = 1.82$ min. The *turnover rate constant* or *fractional turnover rate* of a

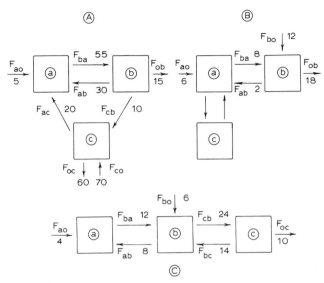

FIG. 24. Models to illustrate dissection of transport rates. Numerical values for individual rates are shown.

pool is the sum of all rate constants of efflux, including those accompanied by reflux in the opposite direction. For pool a it is k_{aa}. It also is turnover rate divided by pool size, i.e., F_{aa}/Q_a. For pool a of size 100 this is 55/100 which is fraction of the pool turned over per unit time. Again, as was true for a single isolated pool, *turnover time* is $1/k_{aa} = 1/0.55 = 1.82$.

DESTINATION OF INPUT AND ORIGIN OF OUTPUT

20. In Model A of Figure 24 part of the material entering via F_{ao} is distributed to output F_{ob}, and part to F_{oc}. Material entering via F_{co} also has a dual distribution. What are the respective distributions? This question is pertinent when the model is viewed as a chemical system comprising molecular species a, b, and c. For example, consider F_{ao} to be a source of carbon to form species a. What proportion of this donation leaves via exit F_{ob}, i.e., in species b, and how much via F_{oc} in c? The calculation is based on a stepwise application of ratios of rates (or corresponding rate constants). First, note that all carbon entering a goes to b. Once in b a certain fraction will leave via F_{ob}. This fraction is the ratio of the output F_{ob} to total output. It is $F_{ob}/(F_{ob} + F_{cb})$, which is 15/25, or 0.6 of all entering material from a. That which moves on via F_{cb} is the fraction $F_{cb}/(F_{cb} + F_{ob})$, which is 10/25. And the fraction of this which is lost via F_{oc} is $F_{oc}/(F_{oc} + F_{ac})$, which is 60/80. Multiplying the latter two fractions together gives 0.3 for the fraction of carbon entering b from a

which leaves via F_{oc}. This means that species a carbon is distributed between F_{ob} and F_{oc} in the ratio 0.6/0.3. In terms of total loss F_{ob} contributes 0.6/0.9, which is 2/3, and F_{oc} contributes 0.3/0.9, which is 1/3. This distribution applies to all species a material, whether immediately from F_{ao}, or recycled via F_{ac} and hence reidentified as a. But the present focus is on that from F_{ao}. In terms of the part of rate F_{ao} (5 mg/min) which shows up at each outlet,

$$\text{via} \quad F_{ob} = (\tfrac{2}{3})(5) = 3\tfrac{1}{3} \text{ mg/min}$$
$$\text{via} \quad F_{oc} = (\tfrac{1}{3})(5) = 1\tfrac{2}{3} \text{ mg/min}$$

Disposal of carbon entering the system via F_{co}, and hence indentified as species c formed via this source, is calculated in similar fashion. Immediate loss via F_{oc} is 60/80, which is 0.75, and that which moves on via F_{ac} is the remaining 0.25. The latter multiplied by fraction out F_{ob} is (0.25)(15/25), which is 0.15. Then,

$$\text{fraction of} \quad F_{co} \quad \text{out} \quad F_{oc} = \frac{0.75}{0.75 + 0.15} = \frac{5}{6}$$

$$\text{fraction of} \quad F_{co} \quad \text{out} \quad F_{ob} = \frac{0.15}{0.75 + 0.15} = \frac{1}{6}$$

and

$$\text{via} \quad F_{oc} = (\tfrac{5}{6})(70) = 58\tfrac{1}{3} \text{ mg/min}$$
$$\text{via} \quad F_{ob} = (\tfrac{1}{6})(70) = 11\tfrac{2}{3} \text{ mg/min}$$

Another relationship is the fraction of each output rate, or fraction of that leaving at a given site, which is derived from F_{ao} and F_{co}, respectively. For example, the fraction of excreted carbon of species b at F_{ob} which came from entrance F_{ao} is $(3\tfrac{1}{3})/(15) = \tfrac{2}{9}$. The model as shown does not give rate constants, however the latter could be used in place of rates to construct multiplying fractions. For example,

$$\frac{F_{ob}}{F_{ob} + F_{cb}} = \frac{F_{ob}/Q_b}{F_{ob}/Q_b + F_{cb}/Q_b} = \frac{k_{ob}}{k_{ob} + k_{cb}} = \frac{k_{ob}}{k_{bb}}$$

Still another computation is for fraction of carbon which *recycles* from one species to another and back again. In the case of species a a certain fraction escapes immediate loss at exits for b and c and returns to a. It is $(\tfrac{10}{25}) \times (\tfrac{20}{80})$, which is $\tfrac{1}{10}$. Thus one-tenth of the donation via F_{ao} returns to pool a. The fraction of F_{co} which returns to c is the same, i.e., $(\tfrac{20}{80})(\tfrac{10}{25})$.

PRODUCTION RATE

21. In Model B of Figure 24 the tracee again will be carbon in a chemical system. Molecular species a arises via a direct donation of preformed molecules containing the carbon in F_{ao}. Species a also arises indirectly via carbon from F_{bo} which carbon picks up other types of atoms not shown (or loses them) in synthesizing a through F_{ab}. Thus, the production rate of species a (PR_a) in terms of donor carbon is F_{ao} plus the fraction of F_{bo} which goes to a. The fraction of species b carbon which goes to a is $F_{ab}/(F_{ab} + F_{ob})$, which is $\frac{2}{20}$. This also is the fraction of F_{bo} which goes to a. Thus, in general form,

$$PR_a = F_{ao} + F_{bo}[F_{ab}/(F_{ab} + F_{ob})] \qquad (15)$$

Specifically, in this model,

$$PR_a = 6 + (12)(\tfrac{2}{20}) = 7.2$$

The rate 7.2 is that for input of new carbon to a, i.e., it is the rate of new appearance of a in terms of carbon, or the rate of first arrival of carbon in species a (exclusive of the rate of movement of recycled carbon). PR_a in Model C may be calculated in two ways. The simplest approach is to consider that the carbon of species b has a *net* rate of loss equal to F_{oc} plus a rate to pool a consisting of F_{ab}. Consequently the fraction to a is $F_{ab}/(F_{ab} + F_{oc})$. Then

$$PR_a = F_{ao} + F_{bo}[F_{ab}/(F_{ab} + F_{oc})] \qquad (16)$$

or

$$PR_a = 4 + (6)(\tfrac{8}{18}) = 6\tfrac{2}{3}$$

A more complex formulation for contribution by F_{bo} may be derived. Recycling of carbon from b to c, back to b, then to a is recognized. Viewing the process in dissected parts, the fraction of b directly to a is $F_{ab}/(F_{ab} + F_{cb})$, or F_{ab}/F_{bb}. Another fraction of b moves to c. It is F_{cb}/F_{bb}. Of this, a subfraction F_{bc}/F_{cc} returns to b, and multiplication again by F_{ab}/F_{bb}, as above, defines an additional subsubfraction to a. Thus, during one cycle the total fraction of species b carbon (or its input F_{bo}) to a is $F_{ab}/F_{bb} + [(F_{cb}/F_{bb})(F_{bc}/F_{cc})](F_{ab}/F_{bb})$. Repeated recycling defines a geometric series of the form

$$\frac{F_{ab}}{F_{bb}} + \frac{F_{ab}}{F_{bb}}\left(\frac{F_{cb}F_{bc}}{F_{bb}F_{cc}}\right) + \frac{F_{ab}}{F_{bb}}\left(\frac{F_{cb}F_{bc}}{F_{bb}F_{cc}}\right)^2 + \cdots$$

The sum of this infinite series is

$$\sum_\infty = \frac{F_{ab}/F_{bb}}{1 - (F_{cb}F_{bc}/F_{bb}F_{cc})} = \frac{F_{ab}F_{cc}}{F_{bb}F_{cc} - F_{cb}F_{bc}} \qquad (17)$$

This is the fraction of F_{bo} which contributes to the production rate of carbon in species a. Substitute numbers and multiply the fraction by 6 and add 4,

$$PR_a = 4 + 6\left[\frac{(8)(24)}{(32)(24) - (24)(14)}\right] = 6\tfrac{2}{3}$$

Dividing each output rate in Eq. (17) by its donor pool size, Q_b or Q_c, converts each F value to a corresponding rate constant, k. In other words, if rate constants are available rather than rates, make each symbol F a symbol k in Eq. (17). Numerically this fraction is $\frac{4}{9}$ in the present example, i.e., the same as $F_{ab}/(F_{ab} + F_{oc})$ in Eq. (16). The fraction has an additional alternate form. Since $F_{cc} = F_{cb}$ in this model, substitute the latter for the former in Eq. (17) and get

$$\text{fraction of carbon } b \text{ to carbon } a = F_{ab}/(F_{bb} - F_{bc})$$

RATE OF IRREVERSIBLE DISPOSAL

22. The production rate of carbon for a particular species, i.e., *new* carbon entering, must be balanced by carbon *irreversibly lost* from this same species. The latter will be called *irreversible disposal rate*, or disposal rate (DR). Thus, for species a, $DR_a = PR_a$, and a determination of PR_a will be that for DR_a also. In addition, DR_a may be calculated by sequential fractions as was the case for PR_a. In Model A of Figure 24 all species a carbon moves to b at a rate of 55. The fraction lost irreversibly via F_{ob} is $F_{ob}/(F_{ob} + F_{cb} + F_{ab})$, which is 15/55. This fraction multiplied by 55 gives a rate of 15 for the portion discharged via F_{ob} (this, of course, is obvious at a glance in this model because 15 is the net outward movement of carbon a which became b). The remainder which moves on via F_{cb} has a rate of 10, which must be multiplied by $F_{oc}/(F_{oc} + F_{ac})$, which is 60/80. This multiplication gives 7.5 for disposal rate at F_{oc}. Adding 15 and 7.5 gives 22.5 for rate of irreversible disposal of species a carbon. This is the same as PR_a which will be $5 + (20/80)(70) = 22.5$. A similar procedure may be applied for DR in any pool in any model. Transfer rate from the original pool to a succession of externally discharging recipient pools is peeled stepwise on the basis of fraction discharged at each step. An apparent paradox is that the sum of disposal rates of separate species may exceed that of total input–output rate for the system as a whole. In Model A the disposal rate of species c carbon is $60 + 20[15/(15 + 10)]$, which is 72. Add this to 22.5 for disposal rate of a and get 94.5 as compared to 75 for total output from the system. This is because some of the carbon was successively shared by pools a and c and was counted for each. From pool a the amount which came to exist also in c, and was ultimately lost via F_{oc}, was 7.5, and the amount originating in c which came to exist in a pending loss via F_{ob} was 12. This is an excess of 19.5 above input–output rate for the system as a whole.

Prediction of an Explicit Solution

23. Prediction is based on a cardinal principle of algebra: the number of unknowns must not exceed the number of independent equations relating the unknowns. To determine all *rate constants* of both interchange and egress of an interchanging three-pool system when one pool is sampled the maximum number of equations is five: the three equations embodying slopes (Eqs. (13)–(15), Appendix II) and any two of the three primary equations for either H, K, or L. If two pools are sampled the maximum number of equations increases to 7, and for all three pools the total increases to 9. These are maximum numbers which, with many models, may not actually be attained. This is because the omission of certain connecting channels may eliminate terms in Eqs. (14), (15), (19), or (20) of Appendix II which are essential in retaining a requisite set of independent simultaneous equations containing the essential parameters. Several examples have already been encountered (paragraphs 15, 17, and 18). Another example would be Model G of Figure 20 if k_{ab} were removed to leave five unknown rate constants (instead of six) to accompany an apparent set of 5 equations. But without k_{ab}, Eqs. (14) and (15) of Appendix II have lost the key constituent $(k_{ab} k_{ba})$ which permits completion of the solution beyond that for k_{aa}.

24.   Determination of *pool sizes* (Q) and *flow rates* (F) is based on three steady-state equations, one of which describes each pool. In unrestricted Model 20A these are

Pool a:

$$F_{ao} + k_{ab} Q_b + k_{ac} Q_c = k_{aa} Q_a$$

Pool b:

$$F_{bo} + k_{ba} Q_a + k_{bc} Q_c = k_{bb} Q_b$$

Pool c:

$$F_{co} + k_{ca} Q_a + k_{cb} Q_b = k_{cc} Q_c$$

Input equals output. The latter also may be broken down into separate output channels, e.g., $k_{aa} Q_a = k_{ba} Q_a + k_{ca} Q_a + k_{oa} Q_a$. With restricted models some of the terms will disappear, but the residual expressions are applicable regardless of which channels or compartments are excluded. Since these working equations cannot exceed the number of pools in the system the number of unknowns calculable is restricted to that same number. Assuming that 3 pools exist and that all rate constants are known, the potential unknowns are Q_a, Q_b, Q_c, F_{ao}, F_{bo}, and F_{co}. Taken as a group, three of these must be known if some have not already been deleted due to some restriction in the system. An overabundance of unknowns still may be consistent with a compromise wherein interrelationships are expressible as ratios. Take Model 20B, paragraph 4. The complete steady-state equations are

Pool a:

$$F_{ao} + k_{ab} Q_b = k_{aa} Q_a$$

Pool b:

$$k_{ba} Q_a + k_{bc} Q_c = k_{bb} Q_b$$

Pool c:

$$k_{cb} Q_b = k_{cc} Q_c$$

From the equation for pool c

$$Q_b/Q_c = k_{cc}/k_{cb} = 0.050/0.082 = 0.61$$

Substitute $0.61Q_c$ for Q_b in the equation for pool b,

$$0.371Q_a + 0.034Q_c = (0.189)(0.61Q_c)$$
$$Q_c/Q_a = 0.371/0.081 = 4.57$$

With pool a as a reference of 1, the size ratios of b to a and c to a are 2.79 and 4.57, respectively.

Prediction of Model

25.   In the models of Figure 20 the SA curves for pools b and c, as shown in Figure 23, were calculated by sampling one pool in a system wherein the pattern of interflow and outflow was known. If all these curves were to be obtained by sampling all three pools rather than calculating any of them, one might suspect that, in reverse manner, the exact configuration of the model could be predicted. Such is not the case. Aside from establishing compatibility and incompatibility the inferences are limited. Note, for example, that the set of 3 SA curves for Models C and F are identical, as are those for Models D and H. Even the identification of a precursor–product relationship is not certain without foreknowledge of the model. To be sure, in Models B, C, and F the occurrence of peak activity in pool c subsequent to that for pool b and with crossover of the curves at the peak for c are features consistent with material in b producing product c. In Models D and H, however, where the curves are quite similar to those of B, C, and F due to the fact that tracer is introduced into pool a in all cases, the late timing of the peak of the c curve does not have this same meaning. Pool c is actually the precursor of a and b since new material enters the system via pool c. In Model I, where c is a side-pool to a, the peak timing of curves b and c depends on the relative size of the corresponding pools. When pool b is the larger (Sol. 1) its curve peaks later than that of c, and vice versa (Sol. 2). Note that the two curves are exactly reversed in the two options (Figure 23). Inferences as to reversibility between specific pools of a system are perhaps the most certain of any which can be made; for example, in Model K the 1-, 2-, and 3-component exponentials of pools a, b, and c, respectively, with successive crossover points at exact peaks of succeeding curves, can mean only a one-way series in line.

CHAPTER 4

COMPARTMENT ANALYSIS: FOUR OR MORE POOLS

Four Pools

1. Included in Appendix II are formulations for the algebraic solution of four-pool models (Eqs. (24)–(35)). General equations for unrestricted interchanging models with five, six, or any number of compartments can be derived. But in working with such equations the algebra becomes forbiddingly complex as the number of pools increases. Even with four pools the process is tedious and laborious. The following example of the solution for a four-pool model will give some notion of this trend.

SOLUTION FOR RATE CONSTANTS

2. The model will be that of Figure 25A. The flow rates and pool sizes which evolve in the calculation are shown. The SA curve obtained by sampling pool a is the solid line in Figure 26. The circles and squares may temporarily be disregarded. With a dose of unity the observed curve has the formula

$$SA_a = 0.4e^{-0.5t} + 0.3e^{-0.2t} + 0.2e^{-0.1t} + 0.1e^{-0.01t}$$

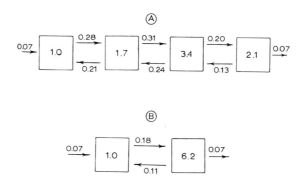

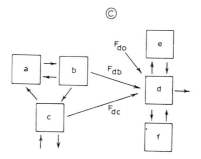

FIG. 25. (A) A four-pool model used for compartment analysis. The SA curve of pool a is shown in Figure 26. (B) A two-pool model for which the SA curve of pool a would generate points denoted by the squares in Fig. 26. (C) Joining two three-pool moieties (see text).

With Q_a known to be 1, the equation for H_1 (Appendix II, Eqs. (28) and (32)) reduces to

$$0.4 = [(-0.5 + k_{bb})(-0.5 + k_{cc})(-0.5 + k_{dd})$$
$$-(-0.5 + k_{bb})k_{cd}k_{dc} - (-0.5 + k_{dd})k_{bc}k_{cb}]/-0.0588 \qquad (1)$$

Similar equations for H_2 and H_3 (Appendix II, Eqs. (28), (33), and (34)) will include g_2 and g_3, respectively, in the numerator rather than g_1, and will have denominators U_2 and U_3, respectively, rather than U_1. Equations for H_2 and H_3 are each subtracted from that for H_1, and the resulting two equations equated to the common group

$$k_{bb}k_{cc} + k_{bb}k_{dd} + k_{cc}k_{dd} - k_{cd}k_{dc} - k_{bc}k_{cb}$$

The remaining halves, when equated to each other, give

$$k_{bb} + k_{cc} + k_{dd} = 0.529 \qquad (2)$$

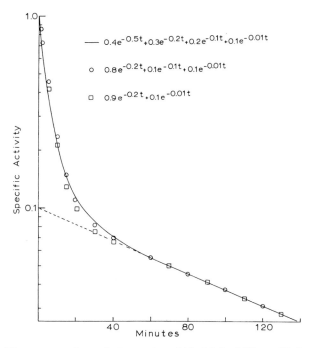

FIG. 26. The curve as drawn is for pool a of Model A of Figure 25. Its equation is indicated. Circles are points on a curve with three-component equation as shown. Squares are for two-component equation as shown, and for Model B of Figure 25.

Since $g_1 + g_2 + g_3 + g_4 = 0.81 = k_{aa} + k_{bb} + k_{cc} + k_{dd}$, this leaves†

$$k_{aa} = 0.281 = k_{ba} \tag{3}$$

Substituting Eq. (2) in that for the difference between the equation for H_1 and that for H_2 (or H_3),

$$k_{bb}k_{cc} + k_{bb}k_{dd} + k_{cc}k_{dd} - k_{cd}k_{dc} - k_{bc}k_{cb} = 0.0644 \tag{4}$$

Equation (25) of Appendix II may be arranged so that it shows the same series of couplets as in Eq. (4) above, leaving k_{aa}, $(k_{bb} + k_{cc} + k_{dd})$, and $-k_{ab}k_{ba}$. The only unknown is k_{ab}, which calculates to be

$$k_{ab} = 0.125 \tag{5}$$

The numerical solutions of Eqs. (3), (4), and (5) now may be substituted for

† Note that k_{aa} also may be obtained directly by adding $H_4 g_4$ to Eq. (18a) of Appendix II and totaling the series. This equation may be extended indefinitely to $H_n g_n$.

corresponding elements which emerge from proper arrangement of Eq. (26) of Appendix II, leaving

$$k_{bb}k_{cc}k_{dd} - k_{dd}k_{bc}k_{cb} - k_{bb}k_{cd}k_{dc} = 0.0117 - (0.281)(0.0644)$$
$$+ (0.281)(0.125)(k_{cc} + k_{dd}) \quad (6)$$

Equation (1), when rearranged, allows substitution of Eqs. (2), (4), and (6) for corresponding groups which appear in it. This leaves

$$k_{cc} + k_{dd} = 0.2234 \quad (7)$$

From Eqs. (2) and (7),

$$k_{bb} = 0.3056 \quad (8)$$

From the model,

$$k_{cb} = k_{bb} - k_{ab} = 0.181 \quad (9)$$

Equation (27) of Appendix II reduces to

$$0.0001 = k_{aa}(k_{bb}k_{cc}k_{dd} - k_{dd}k_{bc}k_{cb} - k_{bb}k_{cd}k_{dc})$$
$$- (k_{cc}k_{dd} - k_{cd}k_{dc})k_{ab}k_{ba} \quad (10)$$

Substituting Eqs. (3), (5), and (6) (with Eq. (7)) into Eq. (10), gives

$$k_{cc}k_{dd} - k_{cd}k_{dc} = 0.00856 \quad (11)$$

Substitute Eq. (7) into Eq. (6) and rearrange,

$$k_{bb}(k_{cc}k_{dd} - k_{cd}k_{dc}) - k_{dd}k_{bc}k_{cb} = 0.00143 \quad (12)$$

Substitute Eqs. (8), (9), and (11) into Eq. (12),

$$k_{dd}k_{bc} = 0.00658 \quad (13)$$

Rearrange Eq. (4),

$$k_{bb}(k_{cc} + k_{dd}) + (k_{cc}k_{dd} - k_{cd}k_{dc}) - k_{bc}k_{cb} = 0.0644 \quad (14)$$

Substitute Eqs. (7), (8), (9), and (11) into Eq. (14),

$$k_{bc} = 0.06872 \quad (15)$$

Substitute Eq. (15) into Eq. (13),

$$k_{dd} = 0.09575 \quad (16)$$

Substitute Eq. (16) into Eq. (7),

$$k_{cc} = 0.12765 \quad (17)$$

From model,

$$k_{dc} = k_{cc} - b_{bc} = 0.05893 \quad (18)$$

Substitute Eqs. (16), (17), and (18) into Eq. (11),

$$k_{cd} = 0.06211$$

From model,

$$k_{od} = k_{dd} - k_{cd} = 0.03364$$

SOLUTION FOR POOL SIZES, FLOW RATES, AND COEFFICIENTS

3. Steady-state equations for three pools are used to complete the solution

$$Q_b k_{bb} = Q_a k_{ba} + Q_c k_{bc} \tag{19}$$

$$Q_c k_{cc} = Q_b k_{cb} + Q_d k_{cd} \tag{20}$$

$$Q_d k_{dd} = Q_c k_{dc} \tag{21}$$

Successive substitutions of Eq. (21) into Eqs. (20) and (19) give

$$Q_d = \frac{Q_c k_{dc}}{k_{dd}}$$

$$Q_c = \frac{Q_b k_{cb} k_{dd}}{(k_{cc} k_{dd} - k_{cd} k_{dc})}$$

$$Q_b = \frac{Q_a k_{ba}(k_{cc} k_{dd} - k_{cd} k_{dc})}{(k_{bb} k_{cc} k_{dd} - k_{bb} k_{cd} k_{dc} - k_{bc} k_{cb} k_{dd})}$$

The values in parentheses need not be calculated again because they were previously determined as conglomerates (Eq. (6) with Eq. (7), and Eq. (11)).

$Q_a = 1$ (independently known), $Q_b = 1.69$, $Q_c = 3.42$, $Q_d = 2.10$

The flow rates emerge from the usual products of rate constants and pool sizes, e.g.,

$$F_{od} = Q_d k_{od} = (2.10)(0.03364) = 0.0706$$

From Eqs. (29)–(35) of Appendix II, the following coefficients are obtained,

	1	2	3	4
K	−0.702	+0.191	+0.295	+0.216
L	+0.349	−0.930	+0.060	+0.520
M	−0.051	+0.525	−0.832	+0.357

Joining Subsystems

4.   Baker *et al.* (1961) employed the six-pool system shown in Figure 25C to represent the transfer of carbon from glucose and its metabolites (pools *a*, *b*, and *c*) to HCO_3^- in pool *d*, which pool in turn is part of a separate three-pool interchanging moiety. Note that the first three-pool moiety transfers carbon irreversibly to *d* via F_{db} and F_{dc}. The analysis of the system depends on labeling species *a* (glucose) with ^{14}C and obtaining a time curve for ^{14}C in *a* (q_a) while also obtaining a companion curve for ^{14}C in HCO_3 (q_d). The latter curve should have six exponential components. As will be noted in subsequent paragraphs, graphic analysis for six components is not likely to be practical. But if pool *d* is labeled separately with $H^{14}CO_3^-$ it should yield only three components in its time curve. The three slopes in this curve will be the same as three of the six to be found in the six-component curve for *d* when *a* is labeled. The other three of the six are the same as those in the three-component curve for *a* when *a* is labeled and sampled directly. Thus the 6 slopes of the six-component curve are predicted to be a combination of these two separately determined groups of three each. Expressed as fraction of dose to *a*, the six-component curve for *d* will be

$$q_d/q_{a0} = M_1 e^{-g_1 t} + M_2 e^{-g_2 t} + M_3 e^{-g_3 t} + M_4 e^{-g_4 t} + M_5 e^{-g_5 t} + M_6 e^{-g_6 t}$$

(22)

The sequence of slopes in this equation has no direct correspondence with respective sequences for the separate three-component curves. For example, the g_2 which goes with M_2 could be the g_3 which goes with K_3 and L_3 of the following paragraph.

5.   Departing from this curve temporarily we may examine the differential equation which defines the fraction of tracer in pool *d* versus time. It is

$$\frac{d(q_d/q_{a0})}{dt} = \frac{k_{db} q_b + k_{dc} q_c + k_{de} q_e + k_{df} q_f - k_{dd} q_d}{q_{a0}}$$

(23)

Now, q_b and q_c in the first two terms are functions of time, the curves for which may be expressed as fraction of the dose added to *a*,

$$q_b/q_{a0} = K_1 e^{-g_1 t} + K_2 e^{-g_2 t} + K_3 e^{-g_3 t}$$

(24a)

$$q_c/q_{a0} = L_1 e^{-g_1 t} + L_2 e^{-g_2 t} + L_3 e^{-g_3 t}$$

(24b)

Substitute Eqs. (24) for q_b and q_c, respectively, into Eq. (23). In a sense these are simultaneously " fed " into *d*. With these substitutions into Eq. (23), it and the conventional differential equations for pools *e* and *f* are three simultaneous equations to be integrated by Laplace transformation and thereby to yield values for the coefficients M_1–M_6 in Eq. (22). These coefficients are expressed in terms of the attending slopes and a variety of rate constants in the system along with coefficients K and L of Eqs. (24). The latter are similarly converted to expressions embodying rate constants and slopes (Appendix II). The resulting expressions for the six M coefficients contain only rate constants and slopes. They are considerably more complex than those for a three-component curve as given in Appendix II. One of the simplest of the six, as an example, is,

$$M_1 = \frac{k_{ba}[k_{db}(k_{cc} - g_1) + k_{dc}k_{cb}](k_{de} - g_1)(k_{df} - g_1)}{(g_2 - g_1)(g_3 - g_1)(g_4 - g_1)(g_5 - g_1)(g_6 - g_1)}$$

These same rate constants appear in the equations for all six coefficients. Values known via the time curve for q_a/q_{a0} are k_{ba} and k_{cc}. In the analysis of this first three-pool moiety, the sum $(k_{oc} + k_{dc})$ had represented one joint output from pool c. And k_{dc} in the six-pool formulations now may be replaced by the following substitution involving this sum

$$k_{dc} = (k_{oc} + k_{dc}) \left(\frac{k_{dc}}{k_{oc} + k_{dc}} \right)$$

In parentheses on the extreme right is the fraction of external output from pool c which goes to d. This, multiplied by total external output, of course, is k_{dc}. The recognition and use of such a product in place of k_{dc} alone will eventually permit solution for both k_{dc} and k_{oc}. The constants k_{db} and k_{cb} may be disposed of by an algebraic maneuver through which they are expressed in terms of k_{oc}, k_{dc}, k_{ac}, k_{cb}, and k_{ab}. The latter two values were defined explicitly by analysis of the first moiety. The constants k_{de} and k_{df} were obtained by analysis of the second moiety. Thus the remaining unknowns are the sum $k_{oc} + k_{dc}$ and the ratio $k_{dc}/(k_{oc} + k_{dc})$. In the observed curve (Eq. (22)) each coefficient resolves to a known numerical value accompanied by the above unknowns. At a chosen point on the curve the values q_d/q_{a0}, t, and the slopes are known. The above unknowns in the coefficients are two in number. The same holds for any *second* point on the curve. These two simultaneous equations yield a solution for k_{oc} and k_{dc}. Solution now is complete because other values for the second three-pool moiety had emerged from calculated rate constants and independent knowledge of the rate F_{od}, or of Q_d. The foregoing development omits many pages of algebra because it is intended only as a survey of principles. In joining two given subsystems the details will vary with the configuration of the model. Although hand calculation is laborious it is less so than for a six-pool model as a whole. Moreover, the slopes and intercepts from observed curves would be more accurately defined than in a six-component curve even if a computer were used to reduce labor.

Errors and False Inferences in Curve Analysis

6.   A source of error in curve analysis previously touched upon in Chapter 2, paragraphs 1 and 25, deserves additional discussion at this point. Because the exact contour of a curve may be difficult to delineate in the face of expected observational errors the exact *number* of separate exponential components which comprise it (and, therefore, values of slopes and intercepts) is not always establishable with confidence. The problem is particularly serious when four or more components are believed to exist. Examples will be given.

7.   The four-component curve analyzed in paragraph 2 is presented as the solid line in Figure 26. Also shown on the graph are a series of open circles representing points on an alternate curve with only three components, for which the formula is

$$SA_a = 0.8e^{-0.2t} + 0.1e^{-0.1t} + 0.1e^{-0.01t}$$

As would be expected the tail of the curve beyond 15 minutes is identical to that of the four-component curve because both curves approach the function $f_4(t) = 0.1e^{-0.01t}$. But even during the first 10 min, when a difference is apparent, a curve fitted to the circles would hardly be distinguishable from that as drawn if observational error were to be tolerated. The squares identify still another curve, in this instance a curve with only two components, the formula of which is shown. That the difference still is not great will be appreciated not only by inspection, but also by noting that the maximum discrepancy with the four-component curve between 15 and 40 min, where correspondence appears poorest, is only 10%. The two-pool model of Figure 25B is an in-line system which would generate such a two-component curve, and which might be invoked as a version wherein pools b, c, and d are consolidated. Note that the reflux to pool a (F_{ab}) is approximately one half that of the four-pool model. The one value for flow rate which differs very little is outflow from the system as a whole. As will be discussed in Chapter 5 this is because the areas subtended by the two curves are very nearly the same.

8.   The solid line of Figure 27A is another example of a four-component curve accompanied by points corresponding to an alternate three-component curve. The formulas are shown. The near identity in contour is striking. The four-component curve is repeated in Figure 27B to illustrate a second possible pitfall. If observations had ceased at 150 min the true tail would have been missed. What was thought to be the final slope between 75 and 150 min (dot–dash line of $f_3(t) = 0.2e^{-0.0077t}$) would, in graphical peeling, yield the three-component curve which fits very nicely up to 150 min (open circles). A precaution which should be observed in any experimental protocol is not only to observe the curve for a long enough span, but also to distribute the separate observations to provide a series of points which define the curve to best

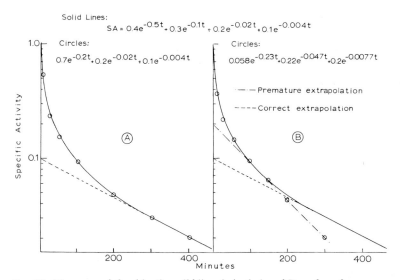

Solid Lines:
$$SA = 0.4e^{-0.5t} + 0.3e^{-0.1t} + 0.2e^{-0.02t} + 0.1e^{-0.004t}$$

Circles:
$$0.7e^{-0.2t} + 0.2e^{-0.02t} + 0.1e^{-0.004t}$$

Circles:
$$0.058e^{-0.23t} + 0.22e^{-0.047t} + 0.2e^{-0.0077t}$$

—·—Premature extrapolation

————Correct extrapolation

FIG. 27. The curves defined by the solid lines in both A and B are for a four-component function having the indicated formula. In part A an alternate three-component curve (formula shown) would be defined by the open circles. In B the consequence of extrapolating a segment preceding the true final slope is shown. The equation accompanying the circles is that which might be predicted if observations were terminated at 150 min and peeling performed on the observed portion of the curve up to that point.

advantage. If the last four hours of sampling are directed toward delineation of the terminal *tail* the measured points during this period might be as far apart as one hour. But at prior points when other components are molding the contour of the curve the sampling must be more frequent. To identify an early rapid slope, the intervals might need to be as short as one minute.

9. Thus it becomes apparent that the contour of a complex exponential curve may be remarkably insensitive to the number of terms in the equation if the exponential constants and coefficients are properly adjusted, especially if observations are made in a limited time span. With the expected errors in data points as obtained from a biologic system the recognition of two vs. three vs. four components may be difficult. To identify five or more components would frequently be impossible. Also note the effect of identical sized pools in an in-line system of nonreversible flow (Chapter 2, paragraphs 21 and 22, and Chapter 7, paragraph 8).

Analysis by Computer

10. Instead of identifying the equation which represents an observed exponential curve and proceding therefrom to derive rate constants by mathematical calculation, a computer can approach the solution in reverse. Rate constants for an assumed model are sought by trial and error until a set is found which will generate a curve matching the one which has been observed. An analog computer consists of a physical system in which a physiologic model is mimicked, whereas a digital computer is essentially a rapid calculator which seeks rate constants providing the best statistical fit to the equation for the observed curve. Computer analysis obviously comes to mind when the number of pools is so great that direct conventional mathematical solution is impractical. The following discussion is limited to underlying principles.

A PHYSICAL ANALOG

11. The analogy of dye washout in a single container of water (Figure 4) may be extended to complex pool systems. The kinetics of a hydrodynamic analog are identical to those of a chemical system. Compartments adjustable to any given capacity and containing constantly stirred fluid might be connected via any assortment of channels fitted with flowmeters through which fluid is propelled by variable-speed pumps. The concentration of dye in a compartment would be monitored with a photometer. By trial and error the pool sizes and flow rates might be adjusted to yield a concentration curve in a given pool which matched that from a counterpart pool of a biological system under study. This would be a tedious and therefore not a very useful approach in practical calculation via an analog system.

12. An electrical device based on the rate of dissipation of charge from a capacitor is the most useful analog for compartment analysis. The basic principle of possible circuitry is fairly simple. Tracer content of a pool (q) is represented by the voltage on a capacitor. In the counterpart of a single pool a pulse charge placed on a capacitor may be permitted to leak across a shunting resistor (outflow from a pool). The fall in potential as it might be monitored on a cathode ray screen will be found to describe a simple exponential decay curve of the form $E = E_0 e^{-(1/RC)t}$ (where E is voltage as a function of time, E_0 is initial voltage, R the resistance of the shunt, and C the capacitance). The fraction $1/RC$ is comparable to the rate constant of flow (k) from a single pool. If, in front of this system, another capacitor is connected serially via a resistor leading from a unidirectional element such as an isolation amplifier in one side of the line, a charge on this first capacitor will leak to the second so as to cause therein a rise in charge and then a fall by virtue of the terminal shunting resistor (exit from the system). This arrangement is comparable to the pool system of Figure 19A. It is apparent that any number of such capacitors may be connected together via resistors to represent flow channels between pools. Each capacitor may be monitored separately, or even simultaneously. Voltage curves may be displayed as rapid iterative traces, as single traces on a long persistence screen, or by a pen writer on moving paper. In operation the value of R or C is varied to create a curve with the form identical to that which has been observed. Appropriate conversion factors for R, C, and t will permit an evaluation of the quantity kt in terms of the electrical equivalents $(1/RC)t$. The value of k is then at hand.

DIGITAL COMPUTER

13. A digital computer calculates the rate constants for a given model by a systematized "guessing" procedure (Berman *et al.*, 1962a and 1962b). In principle the computer is programmed to select rate constants which, after repeated adjustment, approach a set of values which will generate a complex curve within predetermined error criteria, such as least squares, with the best fit to the points which have been obtained in the biologic experiment. In successive trials the constants are adjusted until they converge toward values which approach constancy. The computer also can construct the derived curve in order that it may be compared visually with the series of points which were measured experimentally. Because the searching maneuver would be near endless unless hints were given, rough estimates of at least some of the rate constants should be supplied. Even so, the process is lengthy in comparison to most routine tasks faced by a computer. If prior independent information already has assigned a reliable value to a particular rate constant it may be fixed at the known value and not included among the variables being evaluated for best fit. Or, limits of variation may be assigned to one or more of the rate constants. Such constraint would be very important for a model such as that of Figure 20I, where in sampling pool *a* two explicit solutions exist. Otherwise the convergence might move either to Sol. 1 or Sol. 2.

ADVANTAGES AND LIMITATIONS

14. For the usual two- or three-pool system of known configuration the computer offers no distinct advantage over pencil and paper for calculation of rate constants. Its accuracy is not notably greater than that of graphic analysis. The computer does not "see" much more than the eye even though a least-squares fit, for example, is theoretically more exact. But with models comprising four or more pools, the computer's saving of labor and avoidance of computational error offers a major advantage. Nevertheless the obstacles discussed in paragraphs 6–9 are not circumvented. The addition or subtraction of an exponential component (and the pool responsible for it) may have little effect on the contour of a curve. If such a pool exists the subtleties in contour may be lost in the chance scatter of observed points.

15. The computer can be very useful for *model building*. To be sure, as noted in Chapter 3, paragraph 25, SA curves most often will not prove that a chosen model is the sole correct choice. Nevertheless, if a variety of models are tried for consistency with observed curves, the choice most certainly can be narrowed. Some pools and channels will be proved obligatory whereas others will be dispensable. The rate of transfer in certain channels will be calculable within narrow confidence limits while other rates will be highly uncertain. If it accomplishes nothing else this approach is worthwhile to stimulate a search for unsuspected physiologic mechanisms. An example of such an approach is the study of models for iodine transport made by Berman *et al.*, 1968. Curves of ^{131}I uptake by the thyroid gland, curves of loss from the gland and for labeled hormone from blood, along with excretion rate of iodide in the urine and estimates of body pool sizes for iodide and hormonal iodine were tested for compatibility with a proposed complete model. The effects of adding extra pools, and of deleting or lumping others, were disclosed by generating curves on the computer.

CHAPTER 5

STOCHASTIC ANALYSIS: THE STEWART–HAMILTON EQUATION

Meaning of "Stochastic"

1. "Stochastic" is derived from the Greek word for target. When a series of shots are aimed at a target most will not hit the exact center of the bull's eye. Hence, in a general sense, the word means *conjectural*. Because of the implied random scatter around a central cluster, the word, not surprisingly, has become established in the special vocabulary of probability theory and statistics. A *deterministic* process differs conceptually from a stochastic process in that scatter is nonexistent. Evaluation of a system by compartment analysis is sometimes called a deterministic method, but this means only that the structure of the interior of the system and accompanying internal kinetics are explicitly defined. The movement of individual atoms through discrete pools still is governed by probability. For example, in Figure 22A the fact that 1 % of the dose of tracer is in pool c at 1 min indicates that the chances are 1 in 100 that an atom will have accomplished net forward movement this far, this soon. But what is usually meant by a stochastic method is that compartment analysis is not used. The internal configuration of the system is considered to be unknown. For this reason the system is often dubbed a "black box." Although in the analysis to be presented compartments may be invoked to make concepts more tangible, the working equations are not those of formal compartment analysis.

2.   In kinetic systems labeled with tracer the concept of probability can enter the mathematical treatment in two different ways. Both are based on exceedingly simple concepts: (1) Aside from the influence of nonidentical mass (usually slight), a tracer atom and a natural tracee atom existing beside one another at a given instant have identical chances of appearing at another location in the system after a given interval of time; (2) individual tracer atoms or counterpart tracee atoms comprising a group will not all arrive at some other location or state at the same time; however, there exists a *mean* time for transit of the group as a whole. Although the basic Stewart–Hamilton equation can be derived without invoking these principles, the concepts are uncovered in certain applications. Also, as will be shown in this and the succeeding two chapters, the Stewart–Hamilton equation is closely related to a variety of other formulations which are classified as stochastic. Advantages of "noncompartment analysis" have been emphasized by Meier and Zierler (1954), Tait (1963), Bergner (1962, 1964), Orr and Gillespie (1968), and Bassingthwaighte (1970).

The Stewart–Hamilton Equation

3.   In 1897, Stewart developed a procedure for measuring cardiac output which came to be known as the injection method or dilution method (Stewart, 1897). A known dose of hypertonic saline was infused into the heart of a dog where it became mixed with a quasidiscrete volume of blood. This mixture then moved as a bolus-like mass to be partitioned among arterial branches. Detecting electrodes placed across a major artery signaled the time required for its passage. Pooled blood collected from a companion artery during this same interval was tested for electrical conductivity. By finding the appropriate proportion of predose blood and added saline which gave this same conductivity, the joint volume of blood plus saline propelled by the heart was determined. This volume divided by its time for passage across the electrodes gave rate of cardiac output. In a series of studies beginning in 1929 this so-called injection method was extended and validated in Hamilton's laboratory with dye as tracer (Kinsman *et al.*, 1929). In early experiments serial samples were discharged into tubes on a rotating drum. This provided a plottable time–concentration curve which had a reasonably discrete life. Mean concentration during this interval was calculated simply by averaging a series of concentrations at successive brief intervals. Dose of dye divided by mean concentration gave volume propelled during measured time. Thus, the Stewart–Hamilton equation in its original form might be written

$$\text{rate of volume flow} = \frac{\text{volume of transported bolus of blood}}{\text{time of traversal of bolus across a point}}$$

$$= \frac{\text{dose of tracer/mean concentration of tracer in bolus}}{\text{time of traversal}}$$

The refinements, ramifications, and diverse applications of this principle— even in chemical systems—will provide subject matter for discussion in this and subsequent chapters. First to be considered is the mathematical approach to a time–concentration curve generated by a system receiving a dose of tracer.

Algebraic Derivation

4. So basic is the Stewart–Hamilton equation that its derivation deserves illustration by several methods of approach. The equation not only is a very useful tool; it also has kinship to other stochastic formulations which ultimately will be presented. In Figure 28A the single box symbolizes a

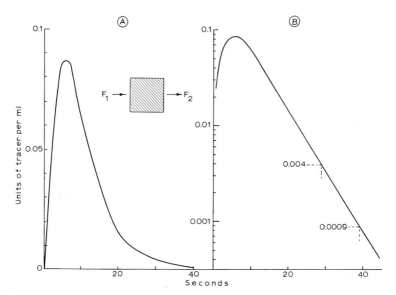

Fig. 28. The box in part A is an undefined complex system yielding the tracer concentration curve at exit as shown on a linear plot. Part B is the same curve on a semilog plot with cut-off points as described in the text.

hydrodynamic system having unknown internal configuration of channels and pools. F_1 and F_2 are, respectively, equal rates of input and output. After a single abrupt delivery of 100 units of tracer at point of input the concentration curve, as shown, is observed via serial or continuous sampling at point of output. The calculation is directed to determining the rate F_2. Assume that the curve to 39 sec (1 % of peak height) represents passage of essentially all tracer. The initial concept is that the tracer, during its 39 sec of existence in the system, was included within a discrete volume of fluid. This volume was not just that of the system itself, but consisted of all participating fluid which moved. Incorporated in it was tracer, which in traversing the point of exit, was present in continuously changing concentration to be monitored in relation to time. If the volume (V) of this fluid which moved in 39 sec (T) could be determined, then

$$F_2 = V/T = V/39 \tag{1}$$

The first step is to determine V. If the unknown volume were that of a static mass of fluid, it could be measured by the simple principle of tracer dilution,

$$V = D/c$$

D is dose of tracer and c is concentration. But because of continual mixing and washout c is not constant during the 39 sec. Therefore, mean concentration must be used in the calculation,

$$V = D/c_{\text{mean}} = 100/c_{\text{mean}} \tag{2}$$

Mean concentration may be determined by measuring the area under the curve in the interval of time from 0–39 sec as by planimetry, and dividing by elapsed time T,†

$$c_{\text{mean}} = A/T = A/39 \tag{3}$$

The area is measured to be 1.11. Thus, for Eq. (3),

$$c_{\text{mean}} = 1.11/39 = 0.0284 \ \text{units/ml}$$

Substituting this value for c_{mean} in Eq. (2),

$$V = 100/0.0284 = 3500 \ \text{ml}$$

---

† The proof evolves by considering the area to be formed by a series of narrow rectangular columns having width Δt and height $c_1, c_2, \ldots, c_n$. The height of each is its area divided by Δt, i.e., $A_1/\Delta t, A_2/\Delta t, \ldots, A_n/\Delta t$. Mean height is the total of all heights divided by the number in the series, and the number in the series is $T/\Delta t$,

$$c_{\text{mean}} = \frac{(A_1 + A_2 + \cdots A_n)/\Delta t}{T/\Delta t} = \frac{A}{T}$$

This becomes exact as width Δt approaches zero.

Thus, in Eq. (1),

$$F_2 = 3500/39 = 90 \text{ ml/sec}$$

Algebraically the foregoing substitutions may be represented as

$$F_2 = \frac{V}{T} = \frac{D/c_{mean}}{T} = \frac{D/(A/T)}{T} = \frac{D}{A}$$

The ratio of dose to area under the time–concentration curve of effluent material is the classical working form of the Stewart–Hamilton equation for rate of flow through a system. A more direct approach is based on the material balance of tracer. The amount introduced (D) is equal to the amount lost. Assuming that essentially all is lost during time span, T,

$$D = F_2 \cdot c_{mean} \cdot T = F_2(A/T)T = F_2 A$$

or

$$F_2 = D/A$$

Derivation by Calculus

5. Note that the time factor has ultimately disappeared in these derivations. This is because the cut-off point at 39 sec was considered to be the time for *essentially all tracer to be cleared* from the system. Actually, if exponential washout from pools exists, the time for complete loss will theoretically be infinite. Infinite time may indeed be recognized in formal treatment by integral calculus. Again take a series of short time spans Δt which are so short that successive instantaneous concentrations of tracer (c_i) may be considered constant during such time. Δt approaches a duration of zero time. Then for each time span, Δt,

$$\text{amount of tracer lost during } \Delta t = F_2 \cdot c_i \cdot \Delta t$$

At infinity all of the dose will be lost, and it will be the sum of the series of foregoing products,

$$D = \lim_{\Delta t \to 0} \sum^{\infty} F_2 \cdot c_i \cdot \Delta t = \int_0^{\infty} F_2 c(t)\, dt$$

If F_2 is assumed *constant*, it may be removed from the integral,

$$D = F_2 \int_0^{\infty} c(t)\, dt$$

or

$$F_2 = \frac{D}{\int_0^{\infty} c(t)\, dt} = \frac{D}{A} \tag{4}$$

In the foregoing form, D is in numerical units (counts, μCi, etc.), and concentration, c, is in numerical units per unit volume. If concentration is expressed as fraction of dose per unit volume, the normalized units on the time
curve are $c(t)/D$, and dose is 1,

$$F_2 = \frac{1}{\int_0^\infty [c(t)/D]\, dt} \tag{4a}$$

Note that this may be construed as a rearrangement of Eq. (4). D, being
constant, may serve either as the principal numerator or may be placed below
in the integral as $1/D$. Because concentration is always measured as a function
of time the symbol (t) will be omitted from subsequent expressions.

ESTIMATE OF AREA

6.   Mathematically, the area in Eq. (4) is measured to infinity. What error is
introduced by cutting it off at 39 sec? That the error is small in the present
instance is easily demonstrated. Figure 28B is a semilog plot of the curve of
Figure 28A. In most biologic systems the terminal tail of such a concentration
curve approximates a simple exponential function, i.e., it is essentially a
straight line on a semilog plot. Assuming that it persists as this simple exponential curve to infinity the residual area is easily calculable. The measured
slope is 0.15. Considering that the residuum begins at 39 sec when the ordinate
value is 0.0009 its equation thereafter is $0.0009e^{-0.15t}$, and the area to infinity
is defined by the integral

$$\int_0^\infty 0.0009e^{-0.15t}\, dt = \frac{0.0009}{0.15} = 0.006$$

This area is insignificant when compared to that of 1.11 for the interval of
time from 0–39 sec. If the curve were cut off at 29 sec, when the ordinate value
at 0.004 would be 5% of peak value, the residual area still would be only
$0.004/0.15 = 0.03$. Thus, the cut-off point is not particularly critical. But if
theoretically maximum accuracy is desired (and the terminal portion of the
curve can with reasonable confidence be predicted to continue at the same
exponential slope) a combination of graphical estimate of area for the main
part of the curve and mathematical integration of the terminal portion is
easily possible by the foregoing approach. Besides planimetry as a graphical
estimate of area, other approaches are to cut out and weigh the paper corresponding to area under a linear plot (weight per unit area being measured),
or to apply the "trapezoid rule." The latter is simply the numerical addition
of separate areas included by a succession of narrow columns, but with the
curve at the top of each column having been reduced to a straight line so as to
form a trapezoid which has an area equal to the product of mean height times

base. With c_1 and c_2 representing heights of sides of a trapezoid, and $t_2 - t_1$ the width of the base,

$$\text{area of trapezoid} = [(c_1 + c_2)/2](t_2 - t_1)$$

Total area is the sum of areas of the series of trapezoids.

Rate of Delivery of Dose

7. Whereas in compartment analysis the mixing of tracer in the primary pool should theoretically be instantaneous, such is not the case when rate is measured by the Stewart–Hamilton equation. That this is true may be appreciated by reasoning as follows: Consider that slow delivery is equivalent to splitting the dose into separate fractions given at successive intervals. For simplicity consider administration in two halves of 50 units each rather than one whole of 100. Delay giving the second dose in order that separate curves can be identified. Obviously, since the height of each curve *at all points* is directly proportional to dose, the area under each curve will be half that for the whole dose. If Figure 28A were so modified each area would be 0.55. If the second curve were shifted so as to begin at zero time, and the ordinate values of the two curves added together, the curve for total dose given at zero time and its subtended area would be reproduced exactly. A rapid succession of very small fractions of the total dose (protracted delivery) can be viewed similarly. Each fraction has the potential to yield a separate curve, but such a group is consolidated in overlap. Each curve need not be lifted back to zero time of onset to show that summation reproduces the subtended area attending a single pulse delivery. The height of each point on the stretched out curve represents a sum of theoretically separate instantaneous values from separate fractions of the dose. The *time* at which these increments to height may occur is irrelevant in relation to their contribution to total area.

Branched System

Mathematics and example

8. Whereas the model of Figure 28A was a system with a single exit, that of Figure 29 has three separate exits. It will be shown that application of the Stewart–Hamilton equation to *any single outlet* will give total inflow–outflow rate, i.e., F_{ao} or the sum, $F_{oa} + F_{ob} + F_{oc}$. Definitive compartments are delineated for purposes of numerical illustration. The previously mentioned concept of probability provides a basis for proof. The fate of unlabeled material (tracee atoms) entering pool a via F_{ao} will be the same as that of tracer atoms of dose D mixed therewith. That is to say, the fraction of tracer lost at any exit will be the same as fraction of inflowing tracee lost at the same point. One

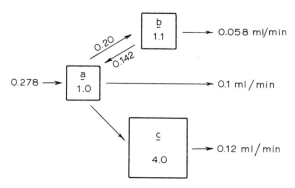

FIG. 29. A branched system to illustrate applicability of the Stewart–Hamilton equation to the SA curve of material moving through any branch from a labeled pool.

essential assumption is that all input is to labeled pool a. When all tracer has cleared the system the quantity of tracer which has left via F_{ob}, for example, will be $\sum_{\infty} F_{ob} \cdot \widetilde{c}_b \cdot \Delta t$. Or,

$$\text{amount of tracer lost via } F_{ob} = F_{ob} \int_0^{\infty} c_b \, dt \qquad (5a)$$

Concentration, c_b, is that in efflux F_{ob} (or that in mixed donor pool b). See paragraph 5 for the comparable expression for amount of tracer lost from a one-exit system. Expressing loss via F_{ob} as fraction of dose lost,

$$\text{fraction of dose lost via } F_{ob} = F_{ob} \int_0^{\infty} c_b \, dt \Big/ D \qquad (5b)$$

The aforementioned principle of probability will require that the fraction of dose of tracer placed in pool a and lost via F_{ob} will be the same as the fraction of inflowing tracee lost via the same route. During any given interval of time the product of entrance or exit rate and time span will give *amount* of tracee moving in or out at any point. For the amount of tracee lost, it is $F_{ob} \cdot T$, and for that entering the system to a it is $F_{ao} \cdot T$. Thus,

$$\text{fraction of entering tracee lost via } F_{ob} = \frac{F_{ob} \cdot T}{F_{ao} \cdot T} = \frac{F_{ob}}{F_{ao}} \qquad (6)$$

Over a *long* period of time the fraction of tracer lost must equal the fraction of entering tracee lost. Equating expressions (5b) and (6),

$$F_{ob}/F_{ao} = F_{ob} \int_0^{\infty} c_b \, dt \Big/ D \qquad (7)$$

Expressing this in terms of F_{ao}

$$F_{ao} = D \Big/ \int_0^{\infty} c_b \, dt$$

For *any* exit, i, from a branched system the same derivation applies. Thus,

$$\text{total input–output rate} = D \Big/ \int_0^\infty c_i \, dt \tag{8}$$

The time–concentration curve, c_i, applies either to effluent material or to the contents of a definitive mixed donor pool.

9. The applicability of Eq. (8) to a branched system can be illustrated numerically by Figure 29. With dose of tracer equal to 1 the equations for concentration as a function of time are

$$c_a = 0.8e^{-0.5t} + 0.2e^{-0.1t}$$
$$c_b = -0.454e^{-0.5t} + 0.454e^{-0.1t}$$
$$c_c = -0.051e^{-0.5t} - 0.086e^{-0.1t} + 0.137e^{-0.03t}$$

Total inflow–outflow rate to the system (Eq. (8)),

Exit a:
$$\frac{1}{(0.8/0.5) + (0.2/0.1)} = \frac{1}{3.6} = 0.278$$

Exit b:
$$\frac{1}{-(0.454/0.5) + (0.454/0.1)} = \frac{1}{3.6} = 0.278$$

Exit c:
$$\frac{1}{-(0.051/0.5) - (0.086/0.1) + (0.137/0.03)} = \frac{1}{3.6} = 0.278$$

The integral to infinity of the time–concentration curve is the same at all exits. Another important fact is noteworthy. Effluent from all or any two outlets may be merged and the mixture sampled for joint concentration. Or, perhaps surprisingly, a given fraction of one may be mixed with a different fraction of another and the concentration curve of the mixture employed for the curve of c_i in Eq. (8). Although provable via algebra the following numerical example will be used instead. Mix one half of output F_{ob} with one third of F_{oc}. To predict the concentration of the mixture the concentrations at separate outlets must be weighted in proportion to their contribution to total. One half F_{ob} is 0.029, and one third F_{oc} is 0.04; total 0.069. Respective weights are $0.029/0.069$, and $0.04/0.069$. Then, from the previously given equations, the concentration curve for the mixture will be the sum of the two after weighting each:

b:
$$\frac{0.029}{0.069} (-0.454e^{-0.5t} + 0.454e^{-0.1t})$$

c:
$$\frac{0.04}{0.069} (-0.051e^{-0.5t} - 0.086e^{-0.1t} + 0.137e^{-0.03t})$$

Sum: $f(t) = -0.221e^{-0.5t} + 0.141e^{-0.1t} + 0.079e^{-0.03t}$

$$D \Big/ \int_0^\infty f(t) \, dt = 1 \Big/ \left(-\frac{0.221}{0.5} + \frac{0.141}{0.1} + \frac{0.079}{0.03} \right) = \frac{1}{3.6} = 0.278$$

Nonexiting pools—"occupancy"

10. In the model just illustrated direct sampling of the pools rather than of their efflux would give the same results. We will turn now to a model where interchanging internal pools not discharging directly to the outside are recognized. All inflow, as before, is into the primary labeled pool. Examples are pool *b* of Models B, F, J, and K, Figure 20, and pool *c* of Models C and I, in the same figure. Here again the time concentration curve from any such pools will give total inflow–outflow for the system as a whole. The proof is similar to that of paragraph 8. In any pool or space within the system to be called V_i, the bit of local flow of *new* material derived from F_{ao} will be called F_i. Therefore, in Eq. (7) replace the left-hand term with F_i/F_{ao} for fraction of input passing through V_i. The quantity of tracer existing at any instant in V_i will be called q_i. Thus the instantaneous concentration is q_i/V_i. The right-hand term of Eq. (7), now represents the fraction of dose *passing through* space V_i. The result is

$$\frac{F_i}{F_{ao}} = \frac{F_i \int_0^\infty (q_i/V_i)\, dt}{D} \tag{9}$$

And F_{ao}, the total input–output rate for the system, is

$$F_{ao} = \frac{D}{\int_0^\infty (q_i/V_i)\, dt}. \tag{10}$$

This is another form of Eq. (8), since $q_i/V_i = c_i$. It may be rearranged, thus

$$F_{ao} = \frac{DV_i}{\int_0^\infty q_i\, dt} \tag{11a}$$

or

$$F_{ao} = \frac{V_i}{\int_0^\infty (q_i/D)\, dt} \tag{11b}$$

Orr and Gillespie (1968) refer to form (11b) as the "occupancy principle." Note that the region V_i is not restricted as to location, connections, size, or homogeneity. The equation says that in any region, the ratio of the capacity, i.e., content of tracee (here volume), to the integral from zero to infinity of the time dependent function describing tracee *content* as fraction of dose is equal to inflow–outflow rate. The ratio is constant for all parts of the system because such ratios being equal to input–output rate are equal to each other. Thus, for two regions V_1 and V_2,

$$\frac{V_1}{\int_0^\infty (q_1/D)\, dt} = \frac{V_2}{\int_0^\infty (q_2/D)\, dt} \tag{12}$$

Equation (12) will be explored further in Chapter 7. It is equally valid for a system wherein capacity of a "region" or "space" is in units of mass, Q, rather than volume, V. Expressions (11) and (12) can be tested and confirmed on the aforementioned models of Figure 20 via equations constructed from Table III.

SPECIAL CASE OF SINGLE POOL

11. In a homogeneous single pool, as described in Chapter 1, paragraph 3, the output, F, is calculable from an integral between *any two* points in time rather than only from zero to infinity. This is because the same slope, k, exists at all times. The numerator of the equation must be adjusted to represent only that portion of dose moving between t_1 and t_2,

$$F = \frac{q_1 - q_2}{\int_{t_1}^{t_2} c\, dt} = \frac{V(c_1 - c_2)}{\int_{t_1}^{t_2} c\, dt} = \frac{V(c_1 - c_2)}{\int_{t_1}^{t_2} c_0 e^{-kt}\, dt} \tag{13}$$

The numerator is equivalent to $Vc_0(e^{-kt_1} - e^{-kt_2})$. The integral in the denominator is $(c_0/k)(e^{-kt_1} - e^{-kt_2})$. This simplifies to $F = kV$, as expected for a one-pool system.

CARDIAC OUTPUT

The model

12. The model of Figure 30A shows right and left ventricles acting as mixing pools separated by lung vessels which are less like a pool than a composite array of parallel tubes of varying length. Nevertheless if tracer is introduced into the right heart the concentration curve at the aorta is not unlike that of the last pool of a three-pool system as shown in Figure 20K. An instantaneous injection into the right ventricle should produce a simple exponential curve which should be remolded and lengthened as a consequence of movement of tracer through an assortment of routes within pulmonary vessels (see Figure 37C), and further reshaped in the left heart pool. The contour of the curve would be affected also by turbulence in large thoracic vessels. But regardless of any combination of compartmental mixing and tubular flow the entire complex can be treated as the " black box " of Figure 28A. Sampling of aortic outflow for concentration vs time would give rate of input–output, i.e., cardiac output. As shown in paragraph 8, a curve obtained from any arterial branch or capillary bed should qualify just as well for the calculation. Also, theoretically, the injection to the right heart need not be instantaneous. It can be delivered non-instantaneously into an arm vein (paragraph 7). But as will be noted below, slow delivery tends to produce a relatively flattened, prolonged primary curve which most likely will be significantly obscured by recirculated tracer.

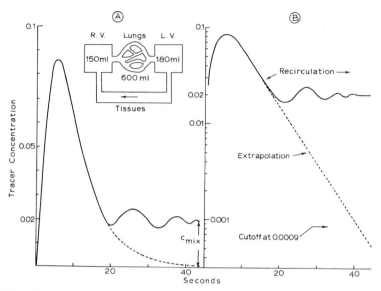

FIG. 30. Part A includes a model for circulation of blood, along with a curve (solid line) for tracer concentration at exit from left ventricle when plot is on linear coordinates. Part B shows extrapolation of the downslope on a semilog plot.

Recirculation

13.   The solid curve of Figure 30A shows that, in contrast to an open system, the downslope of the curve is interrupted by recirculation from peripheral tissues. When total body mixing has been achieved the curve flattens out to a relatively constant value (c_{mix}). In the example shown in Figure 30A the dashed line is what the primary curve might look like if not so interrupted. Figure 30B shows the same curve on a semilog plot with visible primary downslope extrapolated as a straight line. This means that the terminal extension is assumed to be a simple exponential function having the same slope as the portion visualized in the interval between about 12 and 16 sec. If the model were indeed a pure three-pool system the basis for such treatment of the curve would be more secure. The final slope would be a simple exponential term contributed by the largest pool. If the lungs were a true pool containing 600 ml of blood, for example, this slope would be $F/600$. In practice the exact course of the terminal portion of the primary curve is unknown. Nevertheless the logarithmic extrapolation is considered justified by the simple observation that the area thereby delineated gives values for cardiac output which are in reasonable agreement with those estimated by the Fick method. Note that the onset of the curve in Figure 30A has been adjusted to zero time. Bolus flow to a peripheral point would actually delay the onset. The dashed extension in Figure 30A was predicted from points on the extrapolated extension in Figure 30B. The cutoff point for area might be at 39 sec as in Figure 28A and B.

Peripheral monitoring

14. The original technique required serial sampling via arterial puncture and diversion of a stream of blood to a series of tubes on a turntable. The currently popular method is to inject green dye intravenously and record the light absorption curve from a photometer attached to the ear lobe. Conversion of recorder units to blood concentration is by a factor obtained by sampling blood when body mixing is complete (c_{mix}). An alternate calculation of output is by using recorder units to determine fraction of whole body blood volume (BV) which is propelled per unit time, then measuring blood volume to multiply by this fraction. Unless a value for blood volume is assumed this approach does not avoid sampling to evaluate c_{mix}. The fraction of total blood volume propelled is

$$\text{fraction propelled/min} = \frac{\text{vol/min}}{\text{BV}}$$

$$= \frac{D/\int_0^\infty c(t)\, dt}{D/c_{mix}} = \frac{c_{mix}}{\int_0^\infty c\, dt}$$

Insert a common multiplying factor in numerator and denominator to convert blood concentration (c) to the time curve for detector readings (R). These insertions cancel to give an expression in terms of R to be multiplied by BV. Cardiac output is

$$\text{CO} = \text{BV}\, \frac{R_{mix}}{\int_0^\infty R\, dt} = \text{BV}\, \frac{\text{detector after mixing}}{\text{area under detector curve}} \qquad (14)$$

15. An important practical problem is early overriding of the downslope of the primary curve by recirculation. In the present example the direction was predicted by that of a segment only 4 sec in duration. Accurate prediction of this extrapolated segment is obviously sometimes difficult. Early overriding of downslope is minimized by rapid delivery of the dose such as by injecting into the upraised arm or raising the arm rapidly immediately after the intravenous injection. Intracardiac injection and sampling via catheters may serve to sharpen the curve. Another source of error is inconstancy in flow rate associated with the respiratory cycle. Bassingthwaighte *et al.* (1970a, b) have discussed these problems and possible remedies.

Precordial monitoring

16. If a nondiffusing gamma-emitting isotope such as radioiodinated albumin is used as tracer, an activity versus time curve can be recorded directly over the heart. Insofar as the Stewart–Hamilton formula is concerned, the probe need not be aimed at any single chamber or vessel. Paragraphs 8–10 explain its applicability to any combination of fractions of separate sites. But a potential source of error would be the nonuniformity in counting geometry and absorption of radiation, e.g., right ventricle versus left ventricle. Fortunately, existence of a reading after whole body mixing provides a correction. This is readily apparent in Eq. (14). Any attenuation of the area under the primary curve similarly affects the detector reading after mixing. But this means that a wide open uncollimated counter must not be used because considerable chest wall tissue will contribute after mixing but will not contribute to the initial curve itself (Shipley *et al.*, 1953). This component must be subtracted (Donato *et al.*, 1962), or collimation must be relatively narrow over the precordium (Huff *et al.*, 1955).

The Stewart–Hamilton Equation When Rate Is for Mass

17. For the determination of rate of transport or chemical conversion of solutes all previously presented formulas require no change except in notation. Mass, Q (e.g., milligrams), replaces V to represent size or capacity of a pool or "region." Specific activity, SA (e.g., counts per minute per milligram), replaces c. F, for flow rate, is retained but now means rate of movement of unlabeled tracee in terms of mass (e.g., milligrams per minute). Note that in a hydrodynamic system where the rate measured is for fluid, the time curve is for concentration of tracer in fluid, whereas when the rate measured is for mass of solute, the time curve is for concentration of tracer in solute (SA). One of the differences between a biochemical model and that for cardiac output is that tracer is not always introduced at a site where it immediately mixes with all input. As will be shown, the validity of the calculation for input–output rate for a system may be affected by input or output via sites other than for the labeled pool. As was true for compartment analysis (Chapter 2, paragraphs 2 and 27), if major cleavage of participating molecules occurs in chemical transformations the basic calculation must be in terms of tracee *atoms* counterpart to tracer atoms. Also, the molecular species offered as tracer must have label distributed uniformly at all potential sites.

18. As contrasted with a hemodynamic system, the onset of the SA curve for a chemical system is not delayed by a bolus effect. An accurate estimate of the true zero-time value of a curve obtained from the primary labeled pool is not as critical as was noted in Chapter 2 for compartment analysis. A visual extrapolation usually will be adequate if the first observed point is reasonably early. Curve analysis may be performed for purposes of arithmetic integration without undue concern for the number of exponential components. The area is minimally affected by small changes in contour. For example, the curves of Figure 26, when analyzed for 4, 3, or 2 components, give respective areas of 104.3, 105, and 104.5. As was true for compartment analysis a problem may arise in defining the limits of a system under study. Segregation of the kinetics of a chosen system from the effects of communicating connections with other systems in the metabolic milieu of an animal is purely relative. This subject will be taken up in Chapter 8.

LOCATION OF INPUT AND OUTPUT SITES

19. Figure 31 shows an assortment of three-pool models for which all rates and equations for the SA-time curves of separate pools were determined by compartment analysis. The purpose was to apply the Stewart–Hamilton equation to each pool to test its validity for calculating input–output rate for the system as a whole. The equation is

$$\text{Total input–output rate} = D \bigg/ \int_0^\infty \text{SA}_i \, dt \tag{15}$$

where SA_i is the SA-time curve for any given pool. The circular pool *a* is always the labeled pool, and square pools *b* and *c* represent either conversion products of molecular species *a*, or the same species located in other anatomic sites. Near instantaneous mixing occurs only within pools, not between pools. Although in practice the complete configuration of a biologic system may not be known, the sites of labeling, input, and output usually can be identified as discrete pools. If, for a given pool, Eq. (15) gives a correct estimate this pool is left unshaded. If it gives an overestimate or underestimate the pool is shaded according to the code shown. As previously predicted in derivation of the equation and in paragraphs 8–10, *it applies to any pool of a system if all input is into the labeled pool.* If input is elsewhere the *pool of exit invariably qualifies provided it is the sole point of output.* In both these circumstances all new material entering the system can mix with the administered tracer either at the beginning or at the end. If tracee, new to the system, can leave before reaching the labeled pool, as in Model E, the labeled pool does not qualify. Its SA curve is relatively high placed, which means greater area and proportionally underestimated flow rate. The reverse is true for pool *c* in this same model. Note that a secondary side pool is acceptable if attached to a qualifying pool and the side pool has not sequestered tracer (pools *b* and *c*, Models B and C, but not pools *a* and *c*, Model K). In Models L and X, which have two routes of input, a valid estimate cannot be obtained by sampling *any* pool.

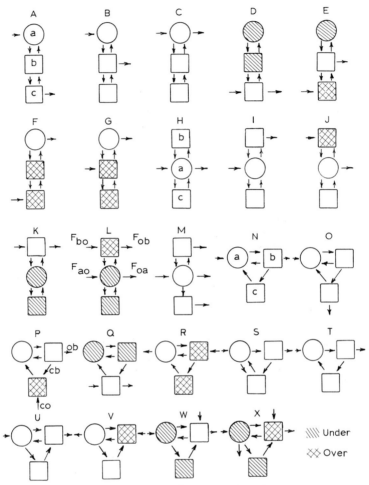

Fɪɢ. 31. Assorted models to illustrate validity of Stewart–Hamilton equation applied at various sites in a system. The circle is always the labeled pool *a*. Pools with single cross hatch give an underestimate of total rate of input–output for the system as a whole; those with double hatch give an overestimate, whereas unmarked pools give a correct result.

CHAPTER 6

STOCHASTIC ANALYSIS: RATES OF PRODUCTION, DISPOSAL, SECRETION, AND CONVERSION; CLEARANCE

Production (or Disposal) Rates of Individual Species

THE PRIMARY LABELED POOL

1. Whereas in Chapter 5 the Stewart–Hamilton equation was employed to estimate the rate of input–output for a system *in toto* (volume or mass), it now will be applied to a system wherein tracee is mass; the rate to be measured will be that for entry or loss of tracee atoms incorporated in a *single* molecular species. The explicit metabolic model of Figure 32A will best explain the principle. Let pools a and b represent dissimilar steroid compounds which exist in blood, and which are metabolically interconvertible. Species a is secreted into blood by testis at a rate of 0.08 mg/min (F_{ao}), and species b is secreted to blood by adrenal at a rate of 0.42 mg/min (F_{bo}). These are respective *secretion* rates. But overall production rate of each species embodies, in addition, another *indirect* source. For species a the total production rate is F_{ao} plus the fraction of F_{bo} which is converted to a rather than excreted via

F_{ob}. Therefore, as explained in Chapter 3, paragraph 21, the production rate of species a is

$$PR_a = 0.08 + 0.42\left(\frac{0.34}{0.34 + 0.38}\right) = 0.28 \text{ mg/min}$$

The Stewart–Hamilton equation, when applied to the SA–time curve of the labeled pool gives production rate for its contained species *regardless of whether all externally derived input comes directly or indirectly into the labeled pool*, or partly via each route as in the model at hand. With a dose of unity the SA curve for labeled pool a in Figure 32A is

$$SA_a = 0.8e^{-0.5t} + 0.2e^{-0.1t}$$

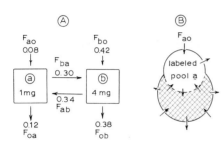

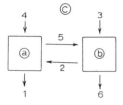

FIG. 32. Models to illustrate calculation of production rate. See text.

The general equation for production rate is

$$PR_a = D\bigg/ \int_0^\infty SA_a \, dt \qquad (1)$$

In the present example, numerical substitution gives

$$PR_a = \frac{1}{(0.8/0.5) + (0.2/0.1)} = 0.28 \text{ mg/min}$$

The applicability of Eq. (1) to any model, regardless of configuration, number of pools, or number and location of sites of input and output, may be confirmed numerically in any system presented in prior chapters. One important assumption is that the system is in steady state, i.e., net production rate of tracee atoms (PR_a) must equal the rate of their irreversible disposal (DR_a),

$$PR_a = DR_a$$

Proof of Eq. (1) emerges from this equality coupled with a similar relationship for tracer. The amount of tracer ultimately lost from the labeled pool must equal the amount introduced (D). Thus, as was shown in Chapter 5, paragraph 5, but with rate of irreversible loss of tracee now being DR_a rather than F_2, the amount of tracer introduced equals its total output,

$$D = \int_0^\infty DR_a \cdot SA_a(t) \, dt$$

or, rearranged with DR_a assumed constant,

$$DR_a = D \bigg/ \int_0^\infty SA_a \, dt \tag{1a}$$

With DR_a and PR_a being equal this leads to Eq. (1). Note that, as expected, PR_a is not the same as total input–output rate for the system as a whole. The latter is 0.5. Equation (1) will give identical values for the two only if all input to, or output from, the system is via the pool which is labeled and sampled.

2. Equation (1) is very useful when the labeled pool is a species contained in blood because most compounds, synthesized or absorbed, enter the blood before disposal. Blood also is usually a pool immediately in line of loss via excretion, utilization, or degradation. But if a compound is partly lost, e.g., utilized, without reaching blood the *overall* production rate for the body as a whole will be underestimated when the calculation is based on the SA curve for blood. For example, let pool a of Figure 32A represent blood glucose receiving input from gut (F_{ao}). Let pool b represent glucose confined in hepatic cells, synthesized via F_{bo}, and partly utilized *in situ* (F_{ob}) in addition to supplying blood glucose via F_{ab}. *Total body* production is F_{ao} plus F_{bo}, which is 0.5. The numerical values used for illustration obviously do not apply to a bona fide animal because hepatic utilization *in situ* is a relatively small fraction of that for the entire body. Equation (1) was mentioned by Tait (1963) for calculation of steroid production rate and was used by Shipley *et al.* (1967) for estimation of glucose disposal rate. The definition of disposal rate, i.e., rate of completely irreversible loss vs loss to sinks, will be discussed in Chapter 8.

A SECONDARY POOL

3. The production–disposal rate of the species in a secondary pool (where label was not introduced) is theoretically calculable in any system, regardless of model, providing the basic equation is adjusted for dose of tracer *actually* *participating* in the sampled pool. For pool b of Figure 32A, the fraction of dose reaching the pool is $0.3/(0.3 + 0.12) = 0.715$. The SA curve for pool b is

$$SA_b = -0.1875e^{-0.5t} + 0.1875e^{-0.1t}$$

Using *fraction* of dose to b rather than full dose of unity (as in the numerator of Eq. (1)),

$$PR_b = \frac{F_{ba}/(F_{ba} + F_{oa})}{\int_0^\infty SA_b(t)\, dt} = DR_b \qquad (2)$$

or, substituting numerical values,

$$PR_b = \frac{0.715}{-(0.1875/0.5) + (0.1875/0.1)} = 0.48 \text{ mg/min}$$

This is confirmed by dissection of flow in the model,

$$PR_b = 0.42 + 0.08\frac{0.3}{0.42} = 0.48 \text{ mg/min}$$

Observe that the sum of PR_a and PR_b is 0.76, which exceeds 0.5, the input–output rate for the system as a whole. This is because a certain fraction of material is counted both for PR_a and PR_b. Equation (2) standing alone has limited practical use since the fraction of dose participating is not ordinarily known. But the equation is useful when combined with a companion expression in double labeling experiments to be described later. Of interest is the effect of substituting certain alternate values for the numerator of Eq. (2) in a model wherein all input is to the labeled pool, e.g., Figure 29. Equation (2), as it stands, gives

$$PR_b = \frac{F_{ba}/F_{aa}}{\int_0^\infty SA_b\, dt}$$

Thus, with the denominator as given in Chapter 5, paragraph 9,

$$PR_b = \frac{0.2/0.42}{3.6} = 0.132$$

The same value is given by dissection of flow,

$$PR_b = 0.2 - 0.142(0.2/0.42) = 0.132$$

As explained in Chapter 5, paragraph 8, the full dose in the numerator gives total input–output rate, which also is PR_a,

$$PR_a = \frac{D}{\int_0^\infty SA_b\, dt} = \frac{1}{3.6} = 0.278 \qquad (2a)$$

However if the numerator consists of the fraction of dose lost via F_{ob} the result is the rate, F_{ob}. Fraction of dose to a which is lost at this point is the same as the ratio of F_{ob} to rate of total loss, i.e., $0.058/0.278 = 0.21$. If tracer output were collected until essentially all was lost from the system, 21% would be recovered at this outlet. Then, with SA as fraction of dose

$$F_{ob} = \frac{\text{fraction of dose lost via } F_{ob}}{\int_0^\infty SA_b \, dt} = \frac{0.21}{3.6} = 0.058 \qquad (2b)$$

Paired with the SA curve from a secondary species (e.g., pool b or its effluent in Figures 32A and 33), a companion curve from labeled pool a will permit a very simple calculation of the fraction of the secondary material (atoms) arising from species a. Employing the symbol i for any secondary species in an unrestricted system,

$$\text{Fraction of } i \text{ arising from } a = \frac{\int_0^\infty SA_i \, dt}{\int_0^\infty SA_a \, dt} \qquad (2c)$$

As usual, 0 to ∞ means a time span which is sufficiently long that most tracer has been processed. Transfer paths to the secondary pool may be direct or indirect via other pools. Take species b of Figure 33B as an example of proof of Eq. (2c). The amount of tracer ultimately carried to b must equal the amount ultimately lost from b. Imagine a group of tracee atoms accompanied by tracer atoms departing from a at a certain net rate to destination b. The amount of tracer therewith ultimately carried to b is the product of rate of movement and the integral of $SA_a(t)$ (Eq. (3)). The counterpart loss of tracer from b is rate of irreversible loss of b atoms multiplied by the integral of their own SA curve. Thus,

$$(\text{Net rate to } b \text{ from } a)\int_0^\infty SA_a \, dt = (\text{Net rate of loss from } b)\int_0^\infty SA_b \, dt$$

In steady state the net rate of loss of atoms from b equals the net rate of total input which includes both atoms from a and atoms not from a. This combined total may be substituted on the right, above, as "total net rate to b." Rearrange the equation and get

$$\frac{\int_0^\infty SA_b \, dt}{\int_0^\infty SA_a \, dt} = \frac{\text{net rate to } b \text{ from } a}{\text{total net rate to } b} = \text{fraction of } b \text{ from } a \qquad (2d)$$

In the model of Figure 32A this is the ratio $F_{ba}/(F_{ba} + F_{bo})$, i.e. F_{ba}/F_{bb}. In such a known model proof readily emerges via the equation for conservation of tracer in the whole system,

$$D = F_{oa}\int_0^\infty SA_a \, dt + F_{ob}\int_0^\infty SA_b \, dt$$

Divide by $\int_0^\infty SA_a \, dt$. The left side becomes DR_a which is $F_{oa} + F_{ba}(F_{ob}/F_{bb})$. In this model the latter ratio is 0.42 which also is the ratio of the integrals, i.e., 1.5/3.6. If excretion rate, F_{ob}, is independently measured to be 0.38 mg/min, a multiplication by 0.42 gives 0.16 mg/min for contribution to F_{ob} by irreversibly disposed a, i.e. the rate of irreversible conversion of a to b. The value of 0.28 for PR_a (via Eq. (1)) minus 0.16 gives 0.12 for F_{oa}. Also, 0.16/0.28 gives 0.57 for fraction of a disposed via F_{ob}. Such calculations are potentially useful for glucose-CO_2 kinetics.

Calculation of Local Output Rate

4. Equation (2b) is actually a special case for a more general equation which gives rate of output from any pool or region in any conceivable model. Proof is as follows: Let F_{oi} represent output rate from region i, and let the total amount of tracer ultimately lost via this channel be $q_{i \, out}$. Then,

$$q_{i \, out} = \sum^\infty F_{oi} \cdot SA_i \cdot \Delta t$$

$$= F_{oi} \int_0^\infty SA_i(t) \, dt \tag{3}$$

$$F_{oi} = q_{i \, out} \Big/ \int_0^\infty SA_i(t) \, dt \tag{4}$$

If a definitive donor pool of size Q_i is recognized, $F_{oi} = Q_i k_{oi}$ and $SA_i = q_i/Q_i$. Substituting into Eq. (3),

$$q_{i \, out} = k_{oi} \int_0^\infty q_i(t) \, dt \tag{5}$$

or

$$k_{oi} = q_{i \, out} \Big/ \int_0^\infty q_i(t) \, dt \tag{5a}$$

Sampling Two Pools

5. In a model such as that represented in Figure 19D with one-way movement from a primary to a secondary pool, both having external egress, sampling pool a and applying Eq. (1) gives F_{ao}. With the integrated time curve for SA_a Eq. (1) becomes

$$F_{ao} = \frac{1}{1/0.03} = 0.03 = F_{oa} + F_{ba}$$

Assume that $F_{ob} = 0.02$ mg/min, a value either independently measured or calculated (by Eq. (4)) by measuring the fraction of dose lost via this outlet over a long period of time while obtaining the time dependent curve of $SA_b(t)$. Since F_{ob} is also DR_b, Eq. (2) gives

$$DR_b = 0.02 = \frac{F_{ba}/0.03}{11.1}$$

$$F_{ba} = 0.0067$$

The remaining rates are calculable by difference. In a more complex system where other pools, reversible or not, are attached to pool a (and may receive input or yield output) the foregoing calculation of F_{ba} still applies. It is also applicable if one-way pools are interposed between a and b. It is not applicable, however, if accompanying F_{oa}, reversible flow to a from b exists, or if external outlets other than that measured exist from b. In such instances F_{ob} would not represent total DR_b. Nonexiting side pools on b are permissible. An application of the foregoing computation might be for production–disposal rate of glucose (F_{ao}) and its rate of conversion to CO_2 (F_{ba}) (Figure 19D). If, in a reversible system such as that in Figure 32A, pools a and b are each labeled *separately* on *separate occasions* Eqs. (1) and (2) may be adapted to give the *fraction of species a converted to b* or vice versa. In one experiment let the dose to a be called D^a, and let the SA–time curve for this pool, when it is the site of labeling be called $SA_a{}^a(t)$. In another experiment place dose D^b in pool b, sample pool a and let the SA curve observed in pool a be designated as $SA_a{}^b(t)$. Then,

$$PR_a = \frac{D^a}{\int_0^\infty SA_a{}^a(t)\,dt} = \frac{D^b F_{ab}/(F_{ab} + F_{ob})}{\int_0^\infty SA_a{}^b(t)\,dt}$$

Equating the two ratios on the right and rearranging gives the fraction of species b which is converted to a,

$$\frac{F_{ab}}{F_{ab} + F_{ob}} = \frac{D^a \int_0^\infty SA_a{}^b(t)\,dt}{D^b \int_0^\infty SA_a{}^a(t)\,dt}$$

The reverse fraction, by a similar approach, is

$$\frac{F_{ba}}{F_{ba} + F_{oa}} = \frac{D^b \int_0^\infty SA_b{}^a(t)\,dt}{D^a \int_0^\infty SA_b{}^b(t)\,dt}$$

When SA values are expressed in terms of fraction of dose the Ds, being 1, may be dropped. See Eq. (7a) and paragraph 12 for an approach which permits complete solution for all rates in the system.

Rate Analysis by SA of Pooled Output

THEORY

6.   Labeled precursor is administered; then an excreted product is collected cumulatively until nearly all tracer has left the system. The calculation of rate, such as production rate of precursor, is based on the SA of the *pooled* collection. If precursor itself is excreted it may be used alternatively for a similar purpose. The duration of collection, although required to be long enough to recover most tracer, must not be so long that SA is too low to measure with reasonable accuracy. A fair compromise might be 95–99% of dose. The recovery is judged on the basis of percent leaving the system, or the percentage of that total *destined to leave* in any one of several products chosen for collection. In the calculations which follow the *duration* of collection will be called T. Assume that Figure 29 represents a precursor a which itself is excreted, and which also yields excreted products b and c. The goal is to determine production rate of a which in this model is the same as that of sole input F_{ao}. Product b will arbitrarily be chosen for collection during time T. The ultimate SA of this pooled collection will be given the symbol α_b. It may be shown as follows that α_b may be used to calculate PR_a in lieu of the time curve for SA_b as employed in Eq. (2a). In Chapter 5, paragraph 4, it was noted that the area, A, under a time–concentration curve between t_0 and a later time t is the same as mean SA during this interval multiplied by the duration of time. Thus, for duration of time, T

$$A = SA_{mean} \cdot T$$

and in the present example, if T is defined as the interval between zero time and t, and SA of product b is measured serially with time, either in a donor pool, or at point of output,

$$A = \text{mean } SA_b \cdot T = \int_0^t SA_b(t)\, dt$$

To be shown is the fact that the foregoing mean SA_b is the same as the SA of pooled product, i.e., α_b. The latter obviously will be

$$\alpha_b = \frac{\text{tracer excreted}}{\text{tracee excreted}} \tag{6}$$

Equation (3), adapted for the finite interval $t = 0$ to t, gives quantity of tracer excreted in time, T, and F_{ob} multiplied by T will give quantity of tracee excreted in the same period. Thus, the SA of pooled collection will be

$$\alpha_b = F_{ob} \int_0^t SA_b(t)\, dt \,/\, F_{ob}T$$

or

$$\alpha_b T = \int_0^t SA_b(t)\, dt$$

Now if T (i.e., the interval from t_0 to t) is sufficiently long to recover nearly all of the tracer the limit t can be called ∞ for practical purposes. Then,

$$\alpha_b T \approx \int_0^\infty SA_b(t)\, dt \tag{7}$$

This, in a strict sense, is an approximation because T cannot be infinite, however it establishes that the product of the SA of pooled product and "long" time of its collection is essentially the same as the classical integral to ∞ of the SA–time curve. This relationship, applied to any pool or species i, may be expressed as a general formula,

$$\alpha_i \cdot T \approx \int_0^\infty SA_i(t)\, dt \tag{7a}$$

In any variant of the Stewart–Hamilton equation the expression αT may be used as an alternate to the integral. In the present example Eq. (2a) becomes

$$PR_a \approx D/(\alpha_b \cdot T) \tag{8}$$

This expression may be converted, via Eq. (6), to an equivalent equation in terms of output of tracer via exit b ($q_{b\,out}$),

$$PR_a \approx D \left/ \frac{q_{b\,out}}{F_{ob} \cdot T}\, T \right. \tag{9a}$$

Regardless of the length of collection (short of actual ∞) the T values cancel; however, T must be long enough that $q_{b\,out}$ is essentially complete. Note another arrangement of Eq. (9a),

$$F_{ob}/PR_a \approx q_{b\,out}/D \tag{9b}$$

The fraction of dose excreted is the same as the fraction of PR_a as represented by F_{ob}. Equation (9b) actually could be written *de novo* because the fraction of dose of tracer which leaves via any branch exit must be the same as that for unlabeled material in a, including new atoms from F_{ao}. Equation (8) then is derivable, in reverse order, by the substitution process denoted in Eq. (9a).

APPLICATION

7. Rate analysis by cumulative excretion of tracer is of particular advantage when the concentration of a compound in blood is so low that a reliable time curve for SA cannot be obtained. The method was first used by Pearlman

(1957) to determine the production rate of progesterone by measuring the cumulative SA of its excreted metabolite. A numerical example for Figure 29 is as follows. Assume product b is collected for 30 min after labeling precursor a with a tracer dose of 1 unit. The predicted SA of pooled output of b is

$$\alpha_b = \frac{0.058\int_0^{30}(-0.454e^{-0.5t}+0.454e^{-0.1t})\,dt}{(0.058)(30)}$$

$$= \frac{0.198}{1.74} = 0.113$$

$$PR_a \approx \frac{1}{(0.113)(30)} = 0.29$$

This estimate, after about 95% recovery of potential output at b compares with the theoretically correct value of 0.278. The same result would emerge via the output of any product. Thus

$$PR_a \approx \frac{D}{\alpha_a T} = \frac{D}{\alpha_b T} = \frac{D}{\alpha_c T} \tag{10}$$

But if a metabolite such as b were derived in part from a source other than a, i.e., if input also existed from outside the system to b, this species could not be used for this purpose. Species a, of course, would still qualify because it is the labeled pool (paragraph 1). And so would c because it, in effect, is an extension of a. An example of a nonqualifying product may be shown numerically for pool b of Figure 32A. Instead of $D/\alpha_b T$, substitute comparable expression $1/\int_0^\infty SA_b\,dt$, as calculated from the equation in paragraph 3. The incorrect result (for PR_a) is 0.67 rather than 0.28. This illustrates the importance of knowing the metabolic model before trusting a particular metabolite to assess production rate of its precursor.

8. An unreasonably short collection time will give an erroneously high approximation of rate. An estimate of completeness of recovery at a given time may be made by plotting cumulative output of tracer and visually judging approach to probable plateau. Also, a correction factor is theoretically applicable if adequate prior experience permits prediction of the fraction of plateau attained at various points in time. This is equivalent to estimating what fraction of ultimately expected total excretion has been attained at earlier points in time. The basis for correction is to convert recovered tracer at too short a time (call it $q_{i\,short}$) to that expected after a long time (call this $q_{i\,out}$). If Eqs. (9a) (and (8)) contains the quantity $q_{i\,short}$ in place of the long term $q_{i\,out}$, multiply the right side by the ratio $q_{i\,short}/q_{i\,out}$. This converts $q_{i\,short}$ to $q_{i\,out}$. Thus, for too short a collection period, T_{short},

$$PR_a \approx \frac{D(q_{i\,short}/q_{i\,out})}{\alpha_i \cdot T_{short}} \tag{11}$$

9. With compounds such as steroids where radiocarbon or tritium is stably incorporated in a part of the molecule which remains intact a convenient formula expresses SA in terms of weight of compound rather than as weight of atomic species. For labeled precursor a and recovered product, i, where $\alpha_i{}^*$ is SA of any pooled product in terms of weight of *compound*,

$$PR_a \approx \frac{(\text{M. Wt.})_a}{(\text{M. Wt.})_i} \frac{D}{\alpha_i{}^*T} \tag{12}$$

TWO POOLS LABELED

The Model

10. For simplicity in development of the theory the definitive two-pool system of Figure 32C will be used. The numbers shown are rates for tracee expressed as mass per unit time. For example, 4 is rate to pool a from the outside (F_{ao}), 2 is rate to a from b (F_{ab}), etc. The problem is to solve for all rates after labeling molecular species a and on another occasion species b, or labeling both species simultaneously with separate isotopes such as ^{14}C and ^3H if both tracer atoms stay with the principal molecular moiety during chemical transformation (e.g., steroid interconversion). The following description will be for the experimental situation in which cumulative output of tracer via both F_{oa} and F_{ob} are measured to give $\alpha_a T$ and $\alpha_b T$; however, as previously explained, in a steady-state system, the integrals of the SA–time curves from the pools could be used for the same purpose. A superscript will be used to denote the species labeled. Thus, $\alpha_b{}^a T$ means label to a while the sample is collected from b.

Equations

11. The basic equations evolve from different expressions for production rate of each species as previously presented. In summary,

| (a)   (b)   (c)   (d) |

$$PR_a \approx \frac{D^a}{\alpha_a{}^a T} = \frac{D^b[F_{ab}/(F_{ab} + F_{ob})]}{\alpha_a{}^b T} = F_{ao} + F_{bo}\left(\frac{F_{ab}}{F_{ab} + F_{ob}}\right) \tag{13}$$

$$PR_b \approx \frac{D^b}{\alpha_b{}^b T} = \frac{D^a[F_{ba}/(F_{ba} + F_{oa})]}{\alpha_b{}^a T} = F_{bo} + F_{ao}\left(\frac{F_{ba}}{F_{ba} + F_{oa}}\right) \tag{14}$$

Column (b) is production rate approximated by sampling the labeled pool or its effluent; column (c) the same rate as given by labeling one pool and sampling the other, and column (d) still another expression for production rate based on dissection of flow.

Solution

12.   Let the experimentally determined values for the system of Figure 32C be

$$\alpha_a{}^aT = \tfrac{4}{19}, \qquad \alpha_a{}^bT = \tfrac{1}{19}, \qquad \alpha_b{}^bT = \tfrac{3}{19}, \qquad \alpha_b{}^aT = \tfrac{5}{38}$$

For purposes of the following illustrative calculations the doses D^a and D^b will be considered unity. In Expression (13) equating of subexpression (a) to subexpression (b) yields: $\mathrm{PR}_a = \tfrac{19}{4}$. This result equated to 13(c) gives $F_{ab}/(F_{ab} + F_{ob}) = \tfrac{1}{4}$. A similar operation in Eq. (14) gives $\mathrm{PR}_b = \tfrac{19}{3}$, and $F_{ba}/(F_{ba} + F_{oa}) = \tfrac{5}{6}$. Note again, in passing, that the ratios $F_{ab}/(F_{ab} + F_{ob})$ and $F_{ba}/(F_{ba} + F_{oa})$ are the respective fractions of b converted to a, and of a converted to b, and that equating columns (b) and (c) might be the prime purpose of an experiment to determine this fraction. Expressions (c) and (d) may now be equated in each row. They form simultaneous equations from which F_{ao} and F_{bo} may be determined. These are 4 and 3, respectively. The next step is via the solution of the following pair of equations describing the steady state of the system,

$$\frac{F_{ab}}{F_{ab} + F_{ob}} = \frac{F_{ab}}{F_{bo} + F_{ba}} \tag{15}$$

$$\frac{F_{ba}}{F_{ba} + F_{oa}} = \frac{F_{ba}}{F_{ao} + F_{ab}} \tag{16}$$

Solution of these simultaneous equations yields $F_{ab} = 2$, and $F_{ba} = 5$. From these results: $F_{oa} = F_{ao} + F_{ab} - F_{ba} = 1$, and $F_{ob} = F_{bo} + F_{ba} - F_{ab} = 6$. The foregoing approach is applicable to more restricted models as in Figures 18 and 19. When a potential channel is absent the corresponding F takes on a value of zero in the equations.

Two precursors and two products

13.   The model of Figure 33A differs from that of Figure 32C in that two products are added. Also, reversible interflow may encounter unknown intermediates (hatched regions). Species a forms product c reversibly. Product c comes not only from a but also indirectly from b. A similar relationship exists on the right in the b, d complex. Equations (13)–(16) remain applicable with the following modifications. Change F_{ob} to F_{od} ; F_{oa} to F_{oc} ; $\alpha_a{}^aT$ to $\alpha_c{}^aT$; $\alpha_a{}^bT$ to $\alpha_c{}^bT$; $\alpha_b{}^bT$ to $\alpha_d{}^bT$ and $\alpha_b{}^aT$ to $\alpha_d{}^aT$. As before, the value of $\int_0^\infty \mathrm{SA}_i(t)\, dt$ can be used in place of $\alpha_i T$ in a steady-state system. To apply such a solution the metabolic model must be known to the extent of appreciating the direct versus indirect source of the products, and also the fact that they have no other outside source.

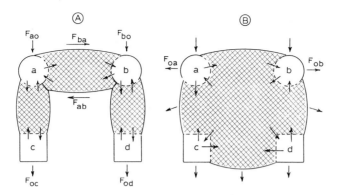

FIG. 33. Partially undefined models to illustrate calculation of rate of externally derived input of two precursors. In Model A each precursor has only one *direct* product. In Model B each precursor can form either product directly without prior conversion to the other precursor.

14. In Figure 33 Model B differs from A in that complete reversibility is allowed between all precursors and products via undefined routes. The basic approach was conceived by Gurpide and co-workers (1965) for assessing secretory rates of separate steroid hormones. Products c and d are collected cumulatively. The principle is based on the probability previously noted, that the fraction of input of tracee to a labeled pool appearing elsewhere in a system will be the same as the fraction of dose of tracer ultimately appearing at this same point. Consider the output F_{oc}. When, during time, T, nearly all tracer has left the system after labeling species a at t_0, let the quantity of tracer recovered here be q_c^a. Then the *fraction* of total dose recovered is q_c^a/D^a. The *total mass* of tracee entering pool a via F_{ao} during this time will be $F_{ao}T$, and the *fraction* of this mass recovered via F_{oc} will be the same as the fraction of dose of tracer recovered. Thus,

$$\text{mass from} \quad F_{ao} \quad \text{to} \quad F_{oc} = (q_c^a/D^a)F_{ao}T$$

The same reasoning applies to input F_{bo} and a separate dose of tracer D^b labeling species b. The fraction of quantity of tracee entering via F_{bo} which appears at F_{oc} is q_c^b/D^b. Thus, the total tracee to F_{oc} from both inputs is

$$\text{total mass to} \quad F_{oc} = (q_c^a/D^a)F_{ao}T + (q_c^b/D^b)F_{bo}T \tag{17}$$

If both sides of Eq. (17) are divided by the total mass to F_{oc} the left side becomes unity and the numerators of the two terms on the right may be rewritten in terms of α, i.e., the SA of the cumulated sample for each respective tracer,

$$1 = (\alpha_c^a T/D^a)F_{ao} + (\alpha_c^b T/D^b)F_{bo} \tag{18}$$

By similar reasoning the equation in terms of specific activities of F_{od} is

$$1 = (\alpha_d{}^a T/D^a)F_{ao} + (\alpha_d{}^b T/D^b)F_{bo} \tag{19}$$

These are two simultaneous equations from which the two unknown input rates F_{ao} and F_{bo} may be determined. Note that, being in terms of SA, they are not affected by procedural losses of compound during purification. If species a and b are sampled directly, their overall production rates are calculable,

$$PR_a \approx D^a/\alpha_a{}^a T \quad \text{and} \quad PR_b \approx D^b/\alpha_b{}^b T$$

Also the fraction of F_{ao} converted to b, or of F_{bo} to a, may be estimated by modification of Eqs. (13d) and (14d) to allow for intermediaries between a and b. Thus, for Eqs. (13a) and (13d),

$$PR_a = F_{ao} + F_{bo} \cdot (\text{fraction of} \quad F_{bo} \quad \text{to} \quad a)$$

The factor in parenthesis includes donations originating in F_{bo} via all conceivable routes. Rearranging,

$$\text{fraction of} \quad F_{bo} \quad \text{to} \quad a = (PR_a - F_{ao})/F_{bo}$$

Augmented models

15. Theoretically, Eqs. (18) and (19) may be extended to fit any model with unlimited species each receiving input, provided each is separately labeled. Let the subscript a represent the first output sampled for α. The superscripts identify the separate labels to successive species receiving input via associated rates F_{ao}, F_{bo}, etc. Then for the first output site,

$$\frac{\alpha_a{}^a T F_{ao}}{D^a} + \frac{\alpha_a{}^b T F_{bo}}{D^b} + \frac{\alpha_a{}^c T F_{co}}{D^c} + \cdots + \frac{\alpha_a{}^n T F_{no}}{D^n} = 1 \tag{20a}$$

For the second,

$$\frac{\alpha_b{}^a T F_{ao}}{D^a} + \frac{\alpha_b{}^b T F_{bo}}{D^b} + \frac{\alpha_b{}^c T F_{co}}{D^c} + \cdots + \frac{\alpha_b{}^n T F_{no}}{D^n} = 1 \tag{20b}$$

and so on until n equations are obtained for n labeled inputs and n outputs,

$$\frac{\alpha_n{}^a T F_{ao}}{D^a} + \frac{\alpha_n{}^b T F_{bo}}{D^b} + \frac{\alpha_n{}^c T F_{co}}{D^c} + \cdots + \frac{\alpha_n{}^n T F_{no}}{D^n} = 1 \tag{20c}$$

If T is the same in all experiments it may be made common to all terms and if SA of pooled samples is expressed as *fraction of dose* per unit weight, all D values are unity. Then, for example, Eq. (20c) becomes

$$(\alpha_n{}^a F_{ao} + \alpha_n{}^b F_{bo} + \alpha_n{}^c F_{co} + \cdots + \alpha_n{}^n F_{no})T = 1 \tag{20d}$$

Serial sampling of output or of donor pools, thereby defining appropriate SA functions, would allow calculation of respective integrated areas, $A_n{}^i$. Equation (20d) then may be restated as

$$A_n{}^a F_{ao} + A_n{}^b F_{bo} + A_n{}^c F_{co} + \cdots + A_n{}^n F_{no} = 1 \qquad (20\text{e})$$

A series of n expressions for n output sites and a corresponding number of input rates consists of a group of simultaneous equations with the number of unknown rates the same as the number of equations. Thus, all input rates are calculable.

Clearance

NET CLEARANCE BY CURVE FOR TRACER VERSUS TIME

16.   In Chapter 1, clearance was described in terms of the kinetics of a single isolated pool. In the present context the principle will apply to a species of solute contained in a delimited volume-space such as blood, however, the species itself will have reversible connections with other compounds in a potentially complex chemical system. An example is species a of Figure 32B. Net clearance of species a is related to its irreversible disposal. By definition, clearance is the volume nominally cleared per unit time. Although no volume is literally cleared, a mathematical equivalent relates the volume which, at the existing concentration of solute, contains an amount of solute corresponding to the quantity irreversibly lost from the specified volume–space (e.g., blood or plasma) per unit time. The *rate of net loss* of species a from such a space is DR_a, its rate of irreversible disposal. Thus, its net clearance, Cl_a, is DR_a divided by its concentration, c_a, expressed as mass per unit volume,

$$Cl_a = DR_a/c_a \qquad (21\text{a})$$

or, rearranging in terms of disposal rate,

$$DR_a = Cl_a \cdot c_a \qquad (21\text{b})$$

17.   The concept of clearance applies not only to unlabeled solute, but also to added tracer. A nominal volume considered entirely cleared of tracee will likewise be entirely cleared of accompanying tracer. Let clearance of tracer be called $Cl_a{}'$. Then,

$$Cl_a{}' = Cl_a$$

The pertinence of this equality is that $Cl_a{}'$ can be obtained by introducing tracer to pool a (labeled species a) and measuring tracer concentration in water (e.g., blood or plasma) versus time. This contrasts with the traditional measurement of clearance by determining solute concentration in blood (denominator in Eq. (21a)) coupled with rate of output obtained by direct chemical measurement of excreted material, e.g., urea clearance by kidney. Measurements are confined to the labeled pool. Let c_a' represent tracer concentration in blood. The rate of loss of tracer will be volume cleared per unit time multiplied by tracer concentration, i.e., $\text{Cl}_a' \cdot c_a'$. But since tracer concentration is not constant, but will decline with time, integral calculus may be invoked in a manner similar to that for the derivation of the equation for DR_a in paragraph 1. The amount lost in any *short* time span, Δt, is the triple product $\text{Cl}_a' \cdot c_a' \cdot \Delta t$. The entire single dose (D) of tracer ultimately lost is the summation of a series of such triple products which, in turn, is the integral from zero to infinity of the time curve $\text{Cl}_a' \cdot c_a'(t)$,

$$D = \int_0^\infty \text{Cl}_a' \cdot c_a'(t)\, dt$$

and if clearance is known to be constant, Cl_a' may be placed outside the integral, leaving behind only the time curve, $c_a'(t)$,

$$D = \text{Cl}_a' \int_0^\infty c_a'(t)\, dt$$

or, rearranging and (because of their equality) substituting clearance of unlabeled solute, Cl_a, for that of tracer, Cl_a',

$$\text{Cl}_a = D \bigg/ \int_0^\infty c_a'(t)\, dt \tag{22}$$

This equation for clearance of species a differs from Eq. (1) (for DR_a) in only one respect. The curve which is integrated is that for concentration of tracer in the *liquid* which jointly harbors tracer and tracee rather than that for SA (which is concentration of tracer *in tracee*). As in numerous previous examples the area under the curve from the time of introduction of the dose to sampled pool a until the time when the value is 1–5% of starting concentration will provide an acceptable value for the integral in Eq. (22). To reiterate: What is being measured is *net* clearance contributed by *all* routes of disposal such as excretion, degradation, and sequestration.

RATE ANALYSIS VIA CLEARANCE

18.   A complete solution for all rates in an interchanging two-pool system (Figure 32A or C), or a solution confined to production rates and fractions of each species converted to the other in incompletely defined models (Figure 33A or B) may be accomplished by employing clearance values coupled with the area under the time curve for concentration of tracer in water (e.g., blood or plasma). Note the following simple relationship between SA, concentration of tracer in water, c', concentration of tracee solute in water, c, and commonly shared volume, V, which harbors tracer in amount q and tracee in amount Q,

$$\text{SA} = \frac{q}{Q} = \frac{q/V}{Q/V} = \frac{c'}{c} \tag{23}$$

Thus, for any species, i, labeled with dose D^i, and existing within a delimited volume, the equivalence in Expression (23) leads to the following modification of Eq. (1),

$$\text{PR}_i = \frac{D^i}{\int_0^\infty [c_i'(t)/c_i]\, dt} \tag{24}$$

and if concentration of solute (c_i) is constant, then,

$$\text{PR}_i = \frac{D^i c_i}{\int_0^\infty c_i'(t)\, dt} \tag{24a}$$

If two interconvertible species both exist in blood the value of volume (V) is common to both; hence both have the same relationship of SA to the ratio of tracer–solute concentration as in Expression (23). This means that SA for *any* species (i) in blood may be expressed as

$$\text{SA}_i = c_i'/c_i \tag{25}$$

Note also that in a steady-state system Eq. (21b) applies similarly to production rate, and that αT is effectively equivalent to $\int_0^\infty \text{SA}(t)\, dt$. Then, choosing the option of retaining D for units of activity in the dose of tracer, and expressing SA as numerical units per unit weight, Expressions (13) and (14) evolve into the following series of equalities:

(a)   (b)   (c)   (d)   (e)

$$\text{PR}_a = \text{Cl}_a c_a = \frac{D^a c_a}{\int_0^\infty c_a'^a\, dt} = \frac{D^b c_a(F_{ab}/F_{bb})}{\int_0^\infty c_a'^b\, dt} = F_{ao} + F_{bo}\left(\frac{F_{ab}}{F_{bb}}\right)$$

$$\text{PR}_b = \text{Cl}_b c_b = \frac{D^b c_b}{\int_0^\infty c_b'^b\, dt} = \frac{D^a c_b(F_{ba}/F_{aa})}{\int_0^\infty c_b'^a\, dt} = F_{bo} + F_{ao}\left(\frac{F_{ba}}{F_{aa}}\right)$$

Solution proceeds in the same manner as outlined in paragraph 12. When c' is expressed as *fraction* of dose per unit volume the symbols D^a and D^b, becoming unity, are dropped.

Limitations

19. The removal of DR from under the integral in the derivation of its fundamental equation in paragraph 1 depends on the assumption that DR is a *constant* rate. An additional assumption for converting DR to PR is that these two are equal. In other words, steady state is assumed for application of Eq. (1) and those expressions which evolve from it. But as will be shown in Chapter 10, Eq. (10) and related expressions embodying αT which follow, are applicable in a certain type of nonsteady state. Rates may change during collection but *all must change proportionally*. For example, in Figure 29 if input rate is doubled *all* output rates must simultaneously be doubled. Equation (10) then will give a weighted mean value for PR_a. But if the relative distribution between F_{oa}, F_{ob}, and F_{oc} changes during collection the result is not a weighted mean for PR_a. This is implicit in Eq. (9b). The fraction of PR_a represented by output rate at a given site is assumed constant. In collecting a steroid metabolite for calculation of production or secretion rate of precursor the collection will require several days. Gallagher *et al.* (1970) have drawn attention to the fact that error is introduced into the estimate of cortisol secretion rate, presumably because this rate is cyclic and the fraction converted to each of an assortment of metabolites varies with its secretion rate.

20.   Equations (20a)–(20e) have certain limitations worth noting. Figure 32C will serve as an illustration of two interconvertible pools receiving input and which jointly might represent a moiety in a larger system. Assume that instead of the relatively slow interchange rates shown these rates (F_{ab} and F_{ba}) are relatively very rapid. In such case this moiety will be indistinguishable from a single pool having a common input rate of 7. F_{ao} and F_{bo} will not be separately identifiable, and $\alpha_a{}^a T$, $\alpha_a{}^b T$, $\alpha_b{}^a T$, and $\alpha_b{}^b T$ will all be approximately 0.14 with unit dose. Input will be about $1/0.14$, i.e., about 7. This, of course, is a practical rather than theoretic limitation. The formulas still apply if values for the product αT are accurate to sufficient significant figures. Ignorance of the true model can invalidate the calculation if inappropriate output sites are sampled. For example in Figure 33B the complete reversibility in all directions makes possible the sampling of *any* two pools, precursor or product. Thus, species c and a could be sampled instead of c and d. The equations would then be

Pool a:

$$\alpha_a{}^a T F_{ao} + \alpha_a{}^b T F_{bo} = 1$$

Pool c:

$$\alpha_c{}^a T F_{ao} + \alpha_c{}^b T F_{bo} = 1$$

But suppose (in a model not shown) that unknown to the observer species c is a compound with *no reversibility* directly or indirectly to b, i.e., situated solely in line of output from a. Then, $\alpha_a{}^a T = \alpha_c{}^a T$, and $\alpha_a{}^b T = \alpha_c{}^b T$. No solution would be possible. Ignorance of the model can lead to error if unrecognized sites of input exist, e.g., into species c and d in the present model. All input sources must be recognized and separately labeled and sampled.

CHAPTER 7

STOCHASTIC ANALYSIS: MEAN TRANSIT TIME, MASS, VOLUME

Concept of Mean Time

1. Although individual tracer atoms have different times of transit from one place or state in a system to another, a *mean* time may be determined for the entire group comprising a dose. Such mean time has equal relevance for tracee atoms. Tracer and tracee, existing side by side, have equal probability of moving to another given site, including external egress. As will be shown, *mean transit time* of tracer from instant of placement in a system to instant of exit, i.e., *mean time for loss*, is related to turnover time for unlabeled material comprising the system. *Mean time of sojourn* of tracer is the same as that for loss because the instant which marks the time of exit is the same as that which completes sojourn. The three terms may be used interchangeably.

Mean Sojourn Time

Sampling Output

Formulas

2. Several working formulas can be devised for calculating mean sojourn time (mean time for loss or transit of tracer). In the following development the internal configuration of the system is *unspecified*, as is the pattern of possible multiple sites of input and output. Although the symbol c will be used for concentration of tracer in moving liquid (tracer per unit volume), the principle is exactly the same for SA (tracer per unit mass). In the second case an output " site " might represent output of a particular molecular species. Assign the symbol F_{oi} to the rate of local output from site i (expressed as volume units per unit time). During an interval of time, Δt, which is sufficiently short that c_i may be considered constant within that interval, the *amount* of tracer lost is equal to the triple product of concentration, output rate, and time span,

$$\text{amount of tracer lost during } \Delta t = F_{oi} \cdot c_i \cdot \Delta t \tag{1}$$

Equation (1) defines a single bit of tracer lost. Adding the series of such bits from t_0 to t_∞ gives the total amount of tracer ultimately lost

$$\text{total tracer lost} = \sum_\infty F_{oi} \cdot c_i \cdot \Delta t \tag{2}$$

Again, consider small "packages" of tracer, each having a different time for *transit*. The transit time for a given small "package" will be called t, i.e., t is the point in time (after entering the system at zero time) when it leaves at site i. This time value is comparable to a *class*, which when multiplied by the *amount* of tracer in the associated bit is comparable to the multiple of class and frequency as encountered in conventional statistics. The multiple here is $t(F_{oi} \cdot c_i \cdot \Delta t)$. This product represents the sum of times for all *individual* units in the bit of tracer. Adding all such sums for all bits then defines a grand sum of times,

$$\text{summation of time values} = \sum_\infty t(F_{oi} \cdot c_i \cdot \Delta t) \tag{3}$$

Conventional calculation of mean time is the sum of times for all units divided by number of units, i.e., Eq. (3) divided by Eq. (2),

$$T_{\text{mean}} = \sum_\infty t(F_{oi} \cdot c_i \cdot \Delta t) \Big/ \Big(\sum_\infty F_{oi} \cdot c_i \cdot \Delta t \Big) \tag{4}$$

Formal integration yields

$$T_{\text{mean}} = \Big(\int_0^\infty t \cdot c_i(t)\, dt \Big) \Big/ \int_0^\infty c_i(t)\, dt \tag{5}$$

The denominator is the area to infinity (or a "long time") under the concentration curve, and the numerator is the area under a derived curve where data points of c_i have been multiplied by the associated time value. If region i is recognizable as a discrete pool having constant size, V, the c_i, above and below, may be multiplied by V (or SA_i multiplied by Q_i for mass) to give an expression in terms of *quantity* of tracer,

$$T_{\text{mean}} = \Big(\int_0^\infty t \cdot q_i(t)\, dt \Big) \Big/ \int_0^\infty q_i(t)\, dt \tag{6}$$

Dividing by D gives an expression in terms of fraction of dose. Or, if a radiation detector or photometer (for dye concentration) views efflux the conversion factor for c_i to detector units (R), being constant, the curve for detector units may be used directly,

$$T_{mean} = \left(\int_0^\infty t \cdot R(t)\, dt\right) \bigg/ \int_0^\infty R(t)\, dt \tag{7}$$

In other words the choice of units is not pertinent as long as they are the same above and below. An important qualification for Eqs. (5)–(7) is that they define mean time *only for that portion of dose of tracer which leaves via the particular exit for which $c_i(t)$ is measured.* As will be demonstrated later, the mean time for the entire system is the sum of *weighted* times for all exits, if multiple sites exist. As previously stated, mean time for loss is the same as mean sojourn time.

Example, single exit present

3. The curves employed for Eqs. (5)–(7) may be of any form and may be integrated by any desired method, but for illustration a multicompartment system yielding a numerically determinable complex exponential curve will be employed (Figure 34A). The integrated form of Eqs. (5)–(7) when the time curve is a complex exponential with n components having intercept coefficients I, and slopes g, is

$$T_{mean} = \frac{(I_1/g_1{}^2) + (I_2/g_2{}^2) + \cdots + (I_n/g_n{}^2)}{(I_1/g_1) + (I_2/g_2) + \cdots + (I_n/g_n)} \tag{8}$$

In the present example the curve for fraction of dose in pool c will be used (Table V)

$$T_{mean} = \frac{+(0.255/0.25) - (1.389/0.01) + (1.134/0.0001)}{+(0.255/0.5) - (1.389/0.1) + (1.134/0.01)} = 112 \text{ min}$$

The value defined by Eq. (8) also is the line parallel to the ordinate upon which the area under a rectilinear plot of the observed curve, when cut out and placed on a straight edge, will balance (equal moments). This is not the same as the line bisecting the bounded area into two equal areas but rather is that of the center of gravity of the plane figure. Note that Eq. (8), if restricted to a single pool, reduces to

$$T_{mean} = \frac{I_1/g_1{}^2}{I_1/g_1} = \frac{1}{g_1} \tag{9}$$

Because the illustrative model is a succession of separate pools which reveal

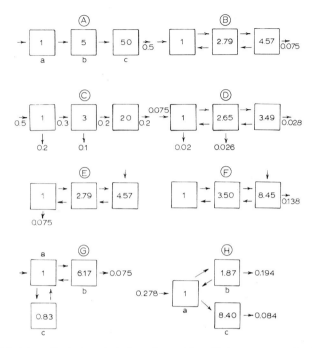

FIG. 34. Models to illustrate calculation of mean time. Rates accompanying the arrows are in units of mass or volume per minute. Numbers in boxes are units of pool size in mass or volume.

their separate slopes in the third pool, the mean times for each pool are calculable and are additive for the system as a whole,

$$T_{mean} = \frac{1}{g_1} + \frac{1}{g_2} + \frac{1}{g_3} \qquad (10)$$

$$= \frac{1}{0.5} + \frac{1}{0.1} + \frac{1}{0.01} = 112 \ min$$

Note from Table V that Eq. (10) also applies for mean tracer sojourn time in an in-line reversible model (Figure 34B). In addition it applies (for sojourn of *tracer*) to Model *F*, where inflow is into the last pool, and would also apply if input were into pool *b* in such a system (not shown). It is not applicable to the other models in this figure.

Example, multiple exits present

4.   As noted in paragraph 2, determination of mean time of sojourn for the dose as a whole in the system as a whole requires that a weighted mean time be calculated for tracer leaving at each exit. Weighting is in terms of fraction

TABLE V

VALUES FOR MODELS OF FIGURE 34[a]

		A	B	C	D	E	F	G	H
H_1	Coefficients for	1	0.7	1	0.7	0.7	0.7	0.7	0.8
H_2	q_a/q_{ao}[b]	—	0.2	—	0.2	0.2	0.2	0.2	0.2
H_3		—	0.1	—	0.1	0.1	0.1	0.1	—
K_1	Coefficients for	−1.25	−0.851	−0.75	−0.807	−0.679	−0.851	−0.300	−0.84
K_2	q_b/q_{ao}	+1.25	+0.511	+0.75	+0.484	+0.408	+0.511	−0.592	+0.84
K_3		—	+0.340	—	+0.323	+0.271	+0.340	+0.892	—
L_1	Coefficients for	+0.255	+0.156	+0.102	+0.119	+0.084	+0.156	−0.408	−0.137
L_2	q_c/q_{ao}	−1.389	−0.850	−0.555	−0.647	−0.458	−0.850	+0.321	−0.187
L_3		+1.134	+0.694	+0.453	+0.528	+0.374	+0.694	+0.087	+0.324
Mean sojourn time, tracer		112	112	48	95.5	76.2	112	107	40.5
Turnover time, system		112	112	48	95.5	112	94	107	40.5

[a] In all models: dose = 1; $g_1 = 0.5$; $g_2 = 0.1$; $g_3 = 0.01$ (when present).

[b] With Q_a being 1.0, these also apply to the SA curve.

of dose lost at each exit. This fraction multiplied by the value obtained by applying Eq. (5), (6), or (7) to an effluent or its donor pool at the corresponding point of output gives weighted mean time for this exit. The sum of all weighted times is mean time for the system as a whole. Model D of Figure 34 will be used as an example. Equation (8) as applied to the slopes and intercepts for respective pools shown in Table V gives unweighted mean time. Divide the respective intercepts by respective pool sizes to get SA in terms of fraction of dose. Then Eq. (3) of Chapter 6 gives fraction of dose lost at each site. The pairs of values and their products are as follows:

	Unweighted mean (min)	Fraction dose lost	Weighted mean time (min)
From a	76.5	0.268	20.5
From b	92.2	0.354	32.7
From c	112.0	0.378	42.3
T_{mean}, whole system			95.5

In an actual experiment all exit sites must be identified and sampled for a time curve. Also, a protracted cumulative collection of tracer would be required to establish the fraction of dose lost.

MEAN TIME BY QUANTITY–TIME CURVE, $q(t)$, ALL POOLS

Formulas

5. If an entire system is viewed as an assortment of subregions wherein tracer units analogous to discrete particles enter and leave each region or subcompartment, including repetitive transit in reversible movements, each unit will have a definitive cumulative time of sojourn *in each region* prior to final exit. For any region, i, a small group comprising n_1 tracer units (approaching in the limit one atom per group) may be assigned a joint sojourn time, T_1. The product of time, T_1 (class), and group population, n_1 (frequency), gives the sum of sojourn times for the units in the group. Another group having n_2 units residing for time T_2 yields a similar product, and so on for many more. The sum of $n_1 T_1 + n_2 T_2 + \cdots + n_n T_n$ is the grand cumulative total of times for separate units. Let n_j and T_j be general symbols for population number and associated time respectively in any group. Then the foregoing grand total is $\sum n_j T_j$, and the conventional calculation for mean time will require that this be divided by total number of units participating,

$$T_{\text{mean}} = \sum n_j T_j \Big/ \sum n_j$$

Each of the separate multiples in the numerator, being a product of units of quantity (ordinate) and time (abscissa), represents a portion of area under the curve for quantity in the region or subcompartment, i, i.e., under the curve $q_i(t)$. Thus the numerator is total area under this curve. Replacing it by the integral,

$$T_{\text{mean}}, \text{region } i = \int_0^\infty q_i(t)\, dt \Big/ \sum n_j$$

The denominator represents *only those units which pass through this region.* In a complex system many will not. In practice the participating quota will not usually be known but a built in weighting factor makes this irrelevant if the ultimate goal is to calculate mean time for the *system as a whole.* If, out of the total dose, D, 75% happens to be involved in region i, then for region i alone $\sum n_j$ is $0.75D$,

$$T_{\text{mean}}, \text{region } i = \int_0^\infty q_i(t)\, dt \Big/ 0.75D$$

But in terms of the entire system the time in this region must be weighted by the fraction of dose so involved. Hence, a multiplication of the foregoing expression by 0.75 gives

$$\text{Region } i, \text{weighted } T_{\text{mean}} = \int_0^\infty q_i(t)\, dt \Big/ D \tag{11}$$

Similar expressions exist for all regions. They are relatively useless when taken *singly* because a *weighted* mean is not the *true* mean for the region. Only when all are added together does a true mean emerge, and it is for the system as a whole. Thus for regions 1, 2, ..., n,

$$T_{\text{mean}}(\text{total system}) = \frac{\int_0^\infty q_1(t)\, dt + \int_0^\infty q_2(t)\, dt + \cdots + \int_0^\infty q_n(t)\, dt}{D} \tag{12}$$

Although Eq. (12) can be applied to contrived multicompartment models, its applicability to a biologic system is unlikely because the system must be divisable into separate discrete compartments and a curve for $q(t)$ obtained for all. But Eq. (12) also may be written

$$T_{\text{mean}} = \int_0^\infty \frac{q_{\text{total system}}(t)\, dt}{D} \tag{13}$$

Note again that Eq. (13) is simply the integral to infinity of the time curve for fraction of dose in the system. It is very useful because the fraction of dose of radioactive tracer in an *entire system* often can be measured with reasonable ease.

Application

6.   Equation (13) is applicable to any system, chemical or hydrodynamic, regardless of whether it consists of an assortment of compartments, channels, labyrinths, or other conceivable components. A continuous gradation in tracer concentration from one part of the system to the other is permissible. Mixing need not be complete in any region. The whole body is a fine example. Whole body counting after giving a gamma-emitting isotope will give a reasonable approximation of the function, $q_{\text{total system}}(t)$, for Eq. (13). Instead of knowing absolute units of tracer present, Eq. (13) may be used with unspecified detector units (R) if readings are referred to that at t_0 when total dose, D, was present, because

$$R(t)/R_0 = q(t)/D$$

Thus,

$$T_{\text{mean}} = \int_0^\infty [R(t)\, dt/R_0]$$

Example of application of Eq. (13)

7.   Let model D, Figure 34, again serve as an example. By totaling the coefficients for the separate pools in Table V an expression is obtained for quantity of tracer in the whole system,

$$\frac{q}{D}(\text{total}) = 0.012e^{-0.5t} + 0.037e^{-0.1t} + 0.951e^{-0.01t}$$

Integration gives

$$T_{mean} = \frac{0.012}{0.5} + \frac{0.037}{0.1} + \frac{0.951}{0.01} = 95.5 \text{ min}$$

A similar approach (or adding the separate integrals for each pool) gives sojourn time for all models in Figure 34 and Table V.

SPECIAL CASE OF EQUAL-SIZED POOLS

8. Appendix II, D gives formulas for the special case where a one-way, in-line model has all pools the same size. (See also Chapter 2, paragraphs 21–23.) Figure 35 shows such a system. From Eq. (9) the mean time for the first pool is $1/0.02 = 50$ min. Each separate pool should have the same mean time because pools are identical in size, are subjected to the same rate of input–output,

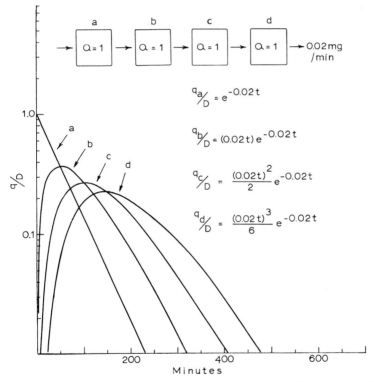

FIG. 35. An in-line, one-way system with equal size pools. Time curves and their equations for fraction of dose, q/D, in each pool are shown.

and process the entire dose of tracer. Accordingly, to satisfy Eq. (12), the successive integrals to infinity of the expressions q_b/D, q_c/D, and q_d/D should each give this same mean time, i.e., $1/0.02$. Integration by Eq. (11) of Appendix V shows that this is indeed true. (See also Eq. (39), Appendix II.) Likewise, the application of Eq. (6) for mean time at point of sole output from the entire system (pool d) gives 200 min total,

$$T_{mean} = \frac{\int_0^\infty t[(0.02t)^3/6]e^{-0.02t}\,dt}{\int_0^\infty [(0.02t)^3/6]e^{-0.02t}\,dt} = \frac{\int_0^\infty t^4 e^{-0.02t}\,dt}{\int_0^\infty t^3 e^{-0.02t}\,dt} = 200$$

The series of curves is notable for the equally spaced (50 min) points of crossover at t_{max}. This means that the derivatives of curves for b, c, and d evaluated at t_{max} (zero slope) should, in sequence be 50, 100, and 150 min. Equation (41) of Appendix II set equal to 0 indeed gives $1/0.02$, $2/0.02$, and $3/0.02$ for the sequence of times for $n = 2$, 3, and 4 successively (pools b, c, and d). (See also Eq. (42), Appendix II.)

All Input to Labeled Pool—Interrelated Formulas

RELATION OF MEAN TRACER TIME AND TURNOVER TIME OF SYSTEM

9.   In any system, regardless of complexity or number of input–output sites, the turnover time for the system is total size (Q_{total} for mass, or V_{total} for volume) divided by total input–output rate. *When all input is into the labeled pool* (pool a, or a comparable region for complete mixing of input and administered tracer), the turnover time and mean time for loss of tracer always are identical. The equation for turnover time is

$$T_{turnover} = Q_{total}/F_{ao} \tag{14}$$

The proof of equivalency follows from Eq. (11b) of Chapter 5. In the derivation of this equation the size of the region V_i was unrestricted within the range of near zero fraction of total volume to that for the entire system. If units of tracee are for mass the comparable equation for the entire system is

$$F_{ao} = \frac{Q_{total}}{\int_0^\infty (q_{total}/D)(t)\,dt} \tag{15}$$

Substituting Eq. (15) for the F_{ao} in Eq. (14),

$$T_{turnover} = \int_0^\infty q_{total}(t)\,dt/D \tag{16}$$

This is the same as Expression (13) for t_{mean}. Thus, for the *whole system, all inflow to labeled pool,*

$$T_{mean} \text{ for tracer} = \text{turnover time for tracee} \qquad (17)$$

10. The obligatory immediate mixing of entering material with tracer in a discrete primary pool or region is not surprising because movement of tracer atoms must reflect movement of all entering nontracer atoms. Were inflow to intrude elsewhere the displacement of unlabeled material and tracer would be dissociated. For example, in Model E of Figure 34 tracer would be lost to the system sooner than tracee in pools *b* and *c*. The reverse would be true for Model F. Note, however, that in all models where inflow is confined to pool *a* the mean time for tracer is the same as turnover time (Table V). Even though a working system may not have pools as discretely formalized as those of Figure 34 the principle still applies.

EQUIVALENCE OF VARIOUS EXPRESSIONS

11. Having demonstrated a variety of equivalencies between equations formulated in this chapter and in Chapters 5 and 6 these expressions now may be brought together to be viewed as a group. They are equivalent only when all input *is to the labeled pool of a system in steady state*. The rates to be shown under (a) at the upper left are equal to each other, and are given by any of the succeeding fractions, (b)–(n). The subscript *i* is a symbol of generality which denotes any pool or delimited region in the system—with one restriction. Two *i* symbols *within* a given fractional expression must represent the same pool or region. For example, Q_i and q_i in (e) are respective quantity of tracee and of tracer in the *same* region. But in another expression such as (f) these symbols may jointly refer to another region. $\sum q_i$ accompanying $\sum Q_i$ means the summation of quantities of tracer in the same combination of regions as for tracee. This summation may be for the entire system. SA is numerical units of tracer divided by associated mass of tracee. In a hydrodynamic system c_i and V_i would be substituted respectively for SA_i and Q_i. The integral $\int_0^\infty$ means subtended area under a curve to a point in time when the ordinate value of a declining curve is a small fraction of its maximum, or mathematical integration to infinity if the tail can, with reasonable confidence, be extrapolated to infinity. Likewise $\alpha_i T$ is the SA of pooled effluent from region *i* when the duration of collection, *T*, is sufficiently long to recover nearly all tracer destined to leave via outlet *i*. The sum $\sum \alpha_i$ is the SA of any combination of pooled outputs collected during the same interval, *T*, from zero time. Dose of tracer is *D* expressed in numerical units (the same as q_{a0}). Parenthetic *t*, as in $q_i(t)$ is the conventional symbol indicating function of time, i.e., time is the independent variable accompanied by q_i or SA_i as dependent variables. PR_a is production rate, species of labeled pool a and DR_a is irreversible disposal rate, species of labeled pool a.

	(a)	(b)

(1) F_{ao}
(2) inflow–outflow rate, entire system
(3) PR_a
(4) DR_a

$$\left.\begin{array}{c}\\ \\ \\ \\ \\ \end{array}\right\} = \dfrac{D}{\int_0^\infty SA_i(t)\,dt}$$

$$\begin{array}{cc} \text{(c)} & \text{(d)} \\[4pt] = \dfrac{D}{\int_0^\infty [q_i(t)/Q_i]\,dt} & = \dfrac{1}{\int_0^\infty [q_i(t)/DQ_i]\,dt} \end{array}$$

$$\begin{array}{cc} \text{(e)} & \text{(f)} \\[4pt] = \dfrac{DQ_i}{\int_0^\infty q_i(t)\,dt} & = \dfrac{Q_i}{\int_0^\infty [q_i(t)/D]\,dt} \end{array}$$

$$\begin{array}{cc} \text{(g)} & \text{(h)} \\[4pt] = \dfrac{D}{\int_0^\infty \left[\sum q_i(t)/\sum Q_i\right]dt} & = \dfrac{D\sum Q_i}{\int_0^\infty \sum q_i(t)\,dt} \end{array}$$

$$\begin{array}{cc} \text{(i)} & \text{(j)} \\[4pt] = \dfrac{\sum Q_i}{\int_0^\infty \sum [q_i(t)/D]\,dt} & = \dfrac{D}{\alpha_i T} \end{array}$$

$$\begin{array}{cc} \text{(k)} & \text{(l)} \\[4pt] = \dfrac{D}{\sum \alpha_i T} & = \dfrac{Q_{\text{total system}}}{\text{mean time of sojourn, total system}} \end{array}$$

$$\begin{array}{cc} \text{(m)} & \text{(n)} \\[4pt] = \dfrac{Q_{\text{total system}}}{\text{turnover time, total system}} & = \dfrac{Q_{\text{total system}}}{\int_0^\infty [R(t)/R_0]\,dt} \end{array}$$

Many of these expressions are interconvertible by simple algebraic rearrangement. For example expression (a2) = expression (b) is the usual form of the Stewart–Hamilton equation, and (c) is the same as (b) but with alternate symbols. The curve in (d) is SA expressed as fraction of dose. The curve for numerical quantity of tracer in (e) is changed in (f) to fraction of dose, leaving Q_i in the numerator. The denominator of expression (f) is called "occupancy" by Orr and Gillespie (1968). Expressions (g)–(i) are for combinations of regions or pools. The remaining examples have been discussed in the foregoing text or in Chapter 6. Expression (n) is for external detector readings when the geometry is such that all parts of the system have equal weight. R_0 is the count rate at peak activity immediately after introduction of the dose, and $R(t)$ represents serial detector readings which constitute the time curve.

CALCULATION OF MASS OR VOLUME

12.   The mass (Q_i) of unlabeled material in a substituent pool or region may be determined by invoking an appropriate combination of the foregoing expressions. Assume that pool a of Figure 34D can be sampled for SA. Its size is unknown. Pool b can be observed for quantity of tracer versus time (as by an externally placed counter), but not sampled for SA. What is the size of pool b? Equate expressions (b) and (f) for pools a and b respectively, remembering that the subscript i means that *any* pool may be chosen for expression (b) and another for (f). Then,

$$\frac{D}{\int_0^\infty SA_a \, dt} = \frac{Q_b}{\int_0^\infty (q_b/D) \, dt}$$

The curves may be predicted from the parameters of Table V. D is unity. Introducing the integrals of the two curves,

$$\frac{1}{13.4} = \frac{Q_b}{35.5}, \qquad Q_b = 2.65 \text{ mg}$$

This principle has been used by Riviere *et al.* (1969) to calculate the quantity of iodine in the thyroid gland. Figure 36 is a simplified version of iodine compartments. The total mass of iodine to be determined is $Q_{(b+c)}$. Administer iodide as a long-lived isotope to label the plasma pool. External counting for a period of months gives $q_{(b+c)}(t)$ for the denominator of Expression (h). Because SA of plasma iodide is not easily measured the best choice is to measure total blood iodine and radioiodine (pools $a + d$). The applicable expression is (g). Thus, equating expressions (h) and (g),

$$\frac{DQ_{(b+c)}}{\int_0^\infty q_{(b+c)} \, dt} = \frac{D}{\int_0^\infty [q_{(a+d)}/Q_{(a+d)}] \, dt} \tag{18}$$

If blood volume is assumed constant during the period of observations an important simplification may be applied to the ratio in the denominator on

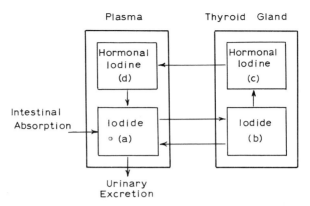

FIG. 36. A simplified model of iodine metabolism to illustrate calculation of mass of iodine in thyroid gland.

the right. Divide both q and Q by blood volume (BV) and get *blood concentration* in terms of liquid,

$$\frac{D}{\int_0^\infty \dfrac{\text{quantity of radioactivity/BV}}{\text{wt. of total iodine/BV}}\, dt}$$

The ratio in the integral is now *concentration* of tracer *per unit volume* of blood divided by concentration of iodine per same unit volume. Iodide plus hormonal iodine are grouped together. If units of volume are called milliliters then the right-hand side of Eq. (18) becomes

$$\frac{D}{\int_0^\infty (\text{radioactivity per milliliter/total stable iodine per milliliter})\, dt}$$

When equated to the left-hand side of Eq. (18), $Q_{(b+c)}$ emerges. Because steady state is assumed, the weight of iodine per unit volume need be determined only occasionally rather than at each sampling for radioactivity.

13. Whole body content of an element such as sodium is calculable in similar fashion. An SA curve for blood is substituted in Expression (b) of paragraph 11. During the same time the denominator of (h) evolves via whole body counting or by sequential subtraction of excreted tracer. Equating these two expressions yields the value of the unknown, $\sum Q_i$, whole body mass of Na. When this method is applied to the whole body an important assumption is that all pools representing total mass are in dynamic exchange with blood. Obviously, the mass of an isolated pool would be excluded from the calculated value. If a quasiisolated pool exchanged very slowly as compared to other pools the curves of blood SA and body content would terminate in prolonged, low-placed, flattened tails. Unless the integrals were obtained over a correspondingly prolonged period the calculated value would exclude the contribution of the slowly exchanging pool.

Special case of pools in line, one-way flow

14.   In the simple one-way model of Figure 34A expression (m) applies to any cumulative abbreviated sequence of pools. Apply Eq. (10): the mean sojourn time for tracer *through* pool a is $1/g = 1/0.5 = 2$ min. *Through* pool b it is $1/0.5 + 1/0.1 = 12$ min. If, for illustration a hydrodynamic system is assumed and flow rate is known to be 0.5 ml/min, the size of the system *through* pool a, i.e., V_a, is $FT = (0.5)(2) = 1$ ml, and *through* pool b, i.e., $V_a + V_b$, it is $(0.5)(12) = 6$ ml. By difference $V_b = 5$ ml. Or, if compartment sampling is not appropriate because a and b are not simple pools, Eq. (5) is applicable to the tracer concentration curves of their effluent material for determination of T, or Eq. (7) to detector readings if the detector has comparable geometry for the two regions. The observed curve in the denominator of these equations and that which is derived from it (numerator) can be integrated by planimetry or any practical method. Again, the equation gives time for all regions prior to and *through* the observed region. An even simpler approach is to plot the observed curve on linear coordinates, cut it out and balance it to determine the center of gravity (paragraph 3). The abscissa value of the point of balance is mean time *through* the observed pool or region. An advantage to direct sampling for tracer concentration rather than using detector units alone is that input–output rate, F, is at the same time calculable by the Stewart–Hamilton equation.

Time for bolus flow versus turnover time of mixing pool

15.   Bolus flow in a hydrodynamic system is comparable to piston displacement. If a piston moves to the end of a cylindrical container so as to expel fluid at a rate of 100 ml/sec and the total volume displaced is 500 ml, the time required must have been 5 sec. Thus,

$$T = V/F \qquad (19)$$

This time, T, is the same as that given by Eq. (14) for turnover time of any "black box" system and for mean time of transit of tracer (expression (a2) and (l) of paragraph 11, with V in place of Q). Thus, the equivalencies of expression (17) may be extended as follows:

(tracee)(tracer)   (tracee)   (tracee or tracer)

$$\frac{V}{F} = T_{\text{mean}} = \text{turnover time} = \text{displacement time by bolus} \qquad (20)$$

The relevance here is that the mean time for passage of tracer through any hybrid combination of pools and tubular channels can be equated to V/F to calculate V if F is known. Except for the bolus mechanism, Eq. (20) applies also to a system with units of mass, where V is replaced with Q.

Examples of hybrid systems

16.   Although the values in Figure 37 are not entirely consonant with normal volumes of blood in two heart chambers separated by lungs, the model will serve to compare the movement of tracer by bolus flow versus exponential washout. In Model A all components are assumed to be mixing pools. Successive turnover times for the three pools (by Eq. (19)) are 1, 5, and 2 sec; total, 8 sec. A concentration curve from pool c will have three slopes: 1, 0.2, 0.5. The sum of their reciprocals is 8 sec for mean sojourn of tracer (Eq. (10)). In Model B the simple exponential curve generated by a moves undisturbed through tube B of 500 ml capacity where it is delayed 5 sec from reference

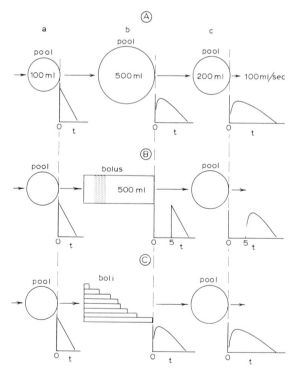

Fig. 37. Comparative effects of exponential washout and bolus propulsion on time-concentration curves. Semilog plot.

point t_0; then in passing c is remolded to form a double exponential curve with slopes 1 and 0.5. The sum of these reciprocals is 3 sec for time from *onset* of the curve which, added to the delay time, gives 8. In Model C the central unit is composed of a series of tubes, which because of their varying length (varying transit times), will remold the curve in approximate resemblance to that caused by a mixing pool. The presence of some very short tubes will allow a few tracer units to appear very promptly in pool c. Therefore the onset of rise is not delayed. The curve for pool c, if plotted on linear coordinates, cut out, and balanced, will give a mean time of 8 sec for the entire system, or Eq. (5) may be used in conjunction with graphical integration of the curves $c_c(t)$ and a derived curve $t \cdot c_c(t)$.

Blood volume in heart and lungs

17. After rapid delivery of radioactive tracer to the right heart a broadly collimated detector over the precordium will record the right heart curve over-ridden on its downslope by the left heart curve. Subtraction of the extrapolated extension of the masked RH curve on a semilog plot will yield a pure LH curve (Shipley *et al.*, 1953). The center of gravity method can be used to determine mean time through each. The difference between the two is mean time through lungs *plus* left heart (not lungs alone). If cardiac output also is determined Eq. (20) gives blood volume of lungs plus left heart. Guintini *et al.* (1963) have calculated pulmonary blood volume separately by invoking a model similar to that of Figure 19F (right and left ventricles) except that a delay caused by a bed of parallel tubes (lung) is interposed between the two pools (as in Figure 37C). The two heart cavities are assumed to be the *same size*. If pulmonary delay did not exist the peak of the left heart curve would be at the time of crossover of the right heart curve (Figure 19F). The shift to the right measured at the ordinate value of the LH peak is pulmonary transit time which when multiplied by cardiac output gives an estimate of the blood volume of lungs. Pulmonary passage is assumed to cause no significant skewing of the concentration curve. An amplitude correction must be made to compensate for differences in tissue absorption and distance of detector to RH as compared to LH.

CHAPTER 8

CLOSED SYSTEMS, CUMULATIVE LOSS, SINKS

1. Curves to be examined in this chapter share a common feature. After an initial rise or fall all of them tend to flatten into a prolonged segment which is horizontal, or nearly so. An example is the curve of activity in the primary labeled pool of a closed system. After an initial decline the curve approaches an absolutely horizontal asymptote. A secondary pool of a closed system first rises then yields a curve which approaches such an asymptote. An open system yields a similar rising curve for cumulative output of tracer. Also, under certain conditions a pool of an open system can generate a curve which, for an extended time, includes a segment which so closely approximates a horizontal plateau that this portion may be treated as an asymptote for certain purposes. Such behavior is characteristic of the curve of SA or quantity of tracer in a pool *much larger* than its fellows. It acts as a sink, i.e., it sequesters entering tracer for such a long time that for short-term phenomena the tracer may be considered irreversibly lost from adjoining pools.

Closed Systems

A THREE-POOL SYSTEM

Equations for activity

2. The most general model for a three-pool closed system is that for Figure 20A with all channels of inflow and outflow removed. Additional selective deletion of some of the channels of interflow will yield a variety of models, three of which are shown in Figure 38A–C. When tracer is introduced into pool *a* of any of these models, or into any variant form having complete interchange between pools, the equations for fraction of dose in each pool as a function of time will have the general form

Pool *a*:

$$q_a/q_{a0} = H_1 e^{-g_1 t} + H_2 e^{-g_2 t} + H_3 \tag{1}$$

Pool *b*:

$$q_b/q_{a0} = K_1 e^{-g_1 t} + K_2 e^{-g_2 t} + K_3 \tag{2}$$

Pool *c*:

$$q_c/q_{a0} = L_1 e^{-g_1 t} + L_2 e^{-g_2 t} + L_3 \tag{3}$$

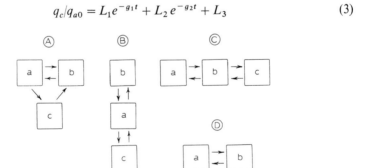

FIG. 38. Assorted models illustrating closed systems (see text).

These are the same as for an open system (Appendix II, Eqs. (4)–(6)), except that the flat terminal slope dictates that $g_3 = 0$, hence the value of $e^{-g_3 t}$ becomes 1. Curves for Model 38A (fraction of dose) are shown in Figure 39A. The terminal plateau has an ordinate value of the final constant term in Eq. (1)–(3). At this point the exponential terms have effectively dropped out, and, for example, $q_a/q_{a0} = H_3$. The equation for fraction of dose in pool *a* for all models of Figure 38 is

$$q_a/q_{a0} = 0.7 e^{-0.5 t} + 0.2 e^{-0.1 t} + 0.1$$

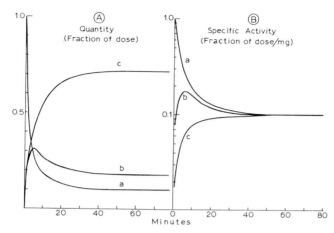

FIG. 39. Part A shows curves for fraction of dose of tracer contained in the three pools of Model A, Figure 38. Part B shows companion SA curves for the same model.

Values for the coefficients of accompanying Eqs. (2) and (3) as they vary among the three models are given in Table VI as are pool sizes. Exponential slopes, g_1 and g_2, are the same in all pools of all models. In Figure 39B are

TABLE VI

VALUES FOR MODELS OF FIGURE 38

	Model A	Model B[a]	Model C
K_1	−0.442	−0.292	−0.840
K_2	+0.262	−0.525	+0.498
K_3	+0.180	+0.817	+0.342
L_1	−0.258	−0.408	+0.140
L_2	−0.462	+0.325	−0.698
L_3	+0.720	+0.083	+0.558
Q_a (mg)	1	1	1
Q_b (mg)	1.80	8.17	3.41
Q_c (mg)	7.21	0.83	5.58
k_{ba}	0.195	0.198	0.370
k_{ab}	0.206	0.024	0.108
k_{cb}	—	—	0.075
k_{bc}	0.024	—	0.046
k_{ca}	0.175	0.172	—
k_{ac}	—	0.206	—

[a] For one solution of a quadratic equation.

curves of SA for Model 38A. The approach to a common asymptote is expected because at equilibrium the probability of tracer atoms existing in any region of the system is the same as that for tracee atoms. Hence everywhere the two must be present in the same ratio. This means an identical SA throughout the system.

Curve analysis

3.   The approach is the same as for an open system. The terminal horizontal tail is extrapolated to the ordinate and subtraction performed. A slight modification is required when, as for pool b in Figure 39A, the curve begins below the asymptote (0.18) and crosses it before approaching it again on the terminal decline. First, subtract the constant asymptotic value, 0.18, from successive points on the observed curve beyond the crossover point at 1.3 min. These values are positive. Next, subtract values on the observed curve from 0.18 prior to the crossover point. These values are negative. Now, make a semilog plot of the positive values only. A straight tail is obtained with slope 0.1, and an extrapolated intercept at 0.262. Finally, subtract values at early points along this extrapolated line from *all* values obtained in the first subtraction, observing signs of the latter. The algebraic difference will become negative at every point to form a straight line the formula for which is $-0.442e^{-0.5t}$. An example for four early points is

t	First subtraction	Final subtraction
0	$0 - 0.180 = -0.180$*	$-0.180 - 0.262 = -0.442$
1	$+0.148 - 0.180 = -0.032$*	$-0.032 - 0.237 = -0.269$
2	$+0.232 - 0.180 = +0.052$	$+0.052 - 0.215 = -0.163$
3	$+0.275 - 0.180 = +0.095$	$+0.095 - 0.194 = -0.099$

Values with asterisks are tabulated but not plotted.

Solution for rate constants and pool sizes

4.   In terms of fraction of dose in each pool the working equations are the same as those for an open system (Eqs. (13)–(23) of Appendix II). The only modification is to make $g_3 = 0$. This means that Eq. (13) is for $g_1 + g_2$, Eq. (14) is for $g_1 g_2$, and Eq. (15) is nonexistent. The denominator in Eqs. (16), (19), and (20) for H_1, K_1, and L_1 becomes $(-g_1 + g_2)(-g_1)$. That for H_2, K_2, and L_2 becomes $(-g_2)(-g_2 + g_1)$. That for H_3, K_3, and L_3 becomes $g_1 g_2$. When any one pool is sampled this series of expressions embodies only 4 independent working equations: two for the intercepts, one for the sum of slopes, and one for the product of slopes. Hence a three-pool model must be restricted to not more than four channels of interchange for explicit evaluation of all rate constants. The solution for these, and for the size of the pools, proceeds as described in Chapters 2 and 3. All of the models in Figure 38 can be solved explicitly by sampling for SA of pool a. As was true for an open system a SA curve from a secondary pool alone is insufficient for complete solution (Chapter 2, paragraph 8). Sampling pool a of Figure 38B gives two explicit solutions (via a quadratic equation which arises in the ensuing algebra) consistent with the symmetry of pools b and c about pool a. This leads to a mathematical interchangeability of b and c. Pool c is designated the smaller in Table VI. If analysis were to begin from this small pool a quadratic form would give an alternate solution not predicted by analysis of the curve for pool a (see Chapter 3 for comparable situation in Table IV). Model C of Figure 38 cannot be solved by sampling pool c because of the singular nature of the attenuation of the equations for coefficients L in terms of rate constants and slopes.

A Two-Pool System

5.   Only one two-pool model is possible (Figure 38D). The expressions for fraction of dose in each pool are
Pool a:

$$q_a/q_{a0} = H_1 e^{-g_1 t} + H_2 \qquad (4)$$

Pool b:

$$q_b/q_{a0} = K_1 e^{-g_1 t} + K_2 \qquad (5)$$

Thus, the curves consist of one exponential term plus a constant. In curve analysis one subtraction gives the exponential term. The equations involving rate constants in relation to slopes and intercepts are now very much simplified. Equation (36) of Appendix II for a two-pool open system becomes, when $g_2 = 0$ and $k_{ab} = k_{bb}$,

$$H_1 = (g_1 - k_{ab})/g_1 \qquad (6)$$

Also from Eq. (18a) of Appendix II, noting that $k_{aa} = k_{ba}$,

$$H_1 = k_{ba}/g_1 \qquad (6a)$$

and, from Eq. (6), substituting $H_2 = 1 - H_1$,

$$H_2 = k_{ab}/g_1 \qquad (7)$$

The equations for the intercepts of the curve for the second pool are similarly modified from those of an open system (Chapter 2, Eq. (7)) to give

$$K_1 = -k_{ba}/g_1 \tag{8}$$

$$K_2 = k_{ba}/g_1 \tag{9}$$

6. Graphic analysis of the SA curve from pool a yields a simple exponential, $I_1 e^{-g_1 t}$, along with I_2, the extrapolated intercept of the asymptote. Then for Eq. (4), $H_1 = I_1/(I_1 + I_2)$ and $H_2 = I_2/(I_1 + I_2)$. Equation (6) or (7) gives k_{ab}, and Eq. (6a) gives k_{ba}. Note that Eq. (13) of Appendix II now says simply that

$$g_1 = k_{ba} + k_{ab} \tag{10}$$

To obtain Q_a from the intercepts of the SA curve (I_1 and I_2), and dose, D,

$$Q_a = D/(I_1 + I_2)$$

Some equivalencies

7. The steady state condition for tracee existing at all times is the basis for the following Eqs. (A)–(D). Equality in vertical direction also exists.

$$\text{(a)} \quad \text{(b)}$$

$$F_{ba} = F_{ab} \tag{A}$$

$$k_{ba} Q_a = k_{ab} Q_b \tag{B}$$

$$Cl_{ba} c_a = Cl_{ab} c_b \tag{C}$$

$$Cl_{ba}(Q_a/V_a) = Cl_{ab}(Q_b/V_b) \tag{D}$$

Cl is clearance, and c is concentration of tracee in water having volume V. After introduced tracer reaches equilibrium its rate of movement now also becomes the same in each direction. After this time (at asymptote),

$$k_{ba} q_{aE} = k_{ab} q_{bE} \tag{E}$$

$$Cl_{ba} c'_{aE} = Cl_{ab} c'_{bE} \tag{F}$$

$$Cl_{ba}(q_{aE}/V_a) = Cl_{ab}(q_{bE}/V_b) \tag{G}$$

The subscript E means at equilibrium, and c' means concentration of tracer in *water*. Equality in vertical direction exists for Eqs. (E), (F), and (G). The foregoing expressions may be rearranged to give certain pertinent ratios. For example,

$$\text{(c)} \quad \text{(d)} \quad \text{(e)} \quad \text{(f)}$$

$$\frac{k_{ba}}{k_{ab}} = \frac{K_2}{H_2} = \frac{Q_b}{Q_a} = \frac{q_{bE}}{q_{aE}} \tag{H}$$

The equality of (c) and (e) comes from Expression (B), and that of (d) and (f) from Eqs. (4) and (5) at infinite time (tracer equilibrium), that of (c) and (d) from Eqs. (7) and (9). Also, from Expressions (C) and (F),

$$\frac{Cl_{ba}}{Cl_{ab}} = \frac{c_b}{c_a} = \frac{c'_{bE}}{c'_{aE}} \tag{I}$$

Note that although SA is the same in both pools at equilibrium the concentration of tracee or tracer in *water* need not be. Expressions (G), (H), and (I) are pertinent in a system where concentrations differ because of a physiologic gradient (Chapter 12, paragraph 3). The foregoing equivalencies may be checked via the following values for Figure 38D when pool *a* receives a unit dose of tracer:

$$SA_a = 0.7e^{-0.5t} + 0.3 \qquad \text{(assigned)}$$
$$SA_b = -0.3e^{-0.5t} + 0.3 \qquad \text{(calculated)}$$
$$Q_a = 1, \qquad Q_b = \tfrac{7}{3} \qquad \text{(calculated)}$$
$$V_a = 5, \qquad V_b = 20 \qquad \text{(assigned)}$$
$$k_{ba} = 0.35, \qquad k_{ab} = 0.15 \qquad \text{(calculated)}$$
$$Cl_{ba} = 1.75, \qquad Cl_{ab} = 3 \qquad \text{(calculated)}$$

Cumulative Loss

SINGLE POOL, SINGLE OUTLET

8. In Model A of Figure 40 pool *a* is a conventional open pool whereas pool *b* is a reservoir which collects effluent. The open top of *b* distinguishes it from a pool of fixed size. It is comparable to a container which can collect unlimited "drainage." Assume that at serial points in time the total quantity of tracer accumulated in *b* is measured after introducing a dose into *a* at zero time. The rising curve will have the general contour of that for pool *c* in Figure 39A. It will rise to an asymptote, but in this case the asymptote will represent recovery of the entire dose, nominally at infinite time. What is the equation for this curve? One simple approach is to consider that the amount in *b* at any given time is the dose offered to *a* minus what remains in *a*. What remains in *a* is the classical equation for decay,

$$q_a = q_{a0}e^{-k_{ba}t} \tag{11}$$

Subtract this from the dose, q_{a0}, placed in *a* at zero time and get an expression for what must be in *b*,

$$q_b = q_{a0} - q_{a0}e^{-k_{ba}t} \tag{12a}$$

or

$$q_b = q_{a0}(1 - e^{-k_{ba}t}) \tag{12b}$$

As fraction of dose in b,

$$q_b/q_{a0} = 1 - e^{-k_{ba}t} \qquad (12c)$$

Equations (12) also may be predicted from Eq. (5) of Chapter 6 which says that the quantity of tracer lost from any pool is the product of rate constant of efflux and the integral of the equation for quantity of dose vs time in the donor pool. In the present case what is lost is in reservoir b. Thus for fraction of dose lost to reservoir b until time, t,

$$q_b/q_{a0} = k_{ba} \int_0^t e^{-k_{ba}t} \, dt \qquad (13a)$$

Integrated,

$$q_b/q_{a0} = (k_{ba}/k_{ba})(1 - e^{-k_{ba}t}) = 1 - e^{-k_{ba}t} \qquad (13b)$$

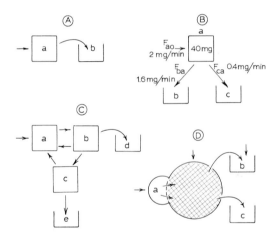

FIG. 40. Models illustrating output to reservoirs (open top containers) which collect tracer and effluent tracee cumulatively.

Assume that such a curve for accumulation is observed experimentally and the goal is to determine the rate constant of output to b from the donor pool, i.e., k_{ba}. The first step is to predict the curve for the donor pool (Eq. (11)). Obviously, a serial subtraction from dose of that which is in b will give what is in a corresponding times. The dose is the value observed at plateau for b. Therefore serial subtraction from the observed plateau gives the equation for q_a. Algebraically, for Eq. (12a), for example, this is

$$q_a = q_{a0} - (q_{a0} - q_{a0} e^{-k_{ba}t}) = q_{a0} e^{-k_{ba}t}$$

This curve of subtraction, on a semilog plot, permits graphic estimation of the slope k_{ba}. If Q_a is independently known, F_{ba} (the same as F_{aa} or F_{ao}) also is calculable.

SINGLE POOL, MULTIPLE OUTLETS

Accumulation curves

9.   Model B of Figure 40 has two cumulating reservoirs, b and c. The problem is to determine all rate constants by obtaining serial cumulative samples of tracer in *one* of these reservoirs, e.g., b. For purposes of illustration the observed curve will be predicted in advance by the relationship given in Eq. (13a). But now the slope for the decay curve of pool a is k_{aa}. Thus,

$$q_b/q_{a0} = k_{ba} \int_0^t e^{-k_{aa}t} dt$$

Integrated,

$$q_b/q_{a0} = (k_{ba}/k_{aa})(1 - e^{-k_{aa}t}) \tag{14}$$

In like manner the equation for accumulation in c is

$$q_c/q_{a0} = (k_{ca}/k_{aa})(1 - e^{-k_{aa}t}) \tag{15}$$

Note that for each reservoir the rate constant (slope) in the exponential term is k_{aa} which is the *total* turnover rate constant for the *donor* pool. The individual rate constants of output appear in the fractional coefficients, each of which represents the ratio of the respective rate constant of local transfer to the overall rate constant of loss from a. Because at infinite time the exponential term is zero, these coefficients are seen to be the asymptotes of the separate cumulation curves for b and c. The curves for all of the pools in Figure 40B are shown in Figure 41A. Their equations are

$$\frac{q_a}{q_{a0}} = e^{-0.05t} \tag{16}$$

$$\frac{q_b}{q_{a0}} = \frac{0.04}{0.05}(1 - e^{-0.05t}) = 0.8 - 0.8e^{-0.05t} \tag{17}$$

$$\frac{q_c}{q_{a0}} = \frac{0.01}{0.05}(1 - e^{-0.05t}) = 0.2 - 0.2e^{-0.05t} \tag{18}$$

The constant terms in Eqs. (17) and (18) are the asymptotes (plateau values). Together they add to the entire dose, and the ratio of each to their sum is the fraction of dose to the respective outlets. Their ratio to each other is the same as the ratio of the respective rate constants or rates,

$$k_{ba}/k_{ca} = F_{ba}/F_{ca} = 0.8/0.2$$

Curve analysis

10. Assume that the values for the model of Figure 40B are unknown and that experimental observations yield the curve of Figure 41A for accumulation of tracer in reservoir *c*. Note that when observations are terminated at 70 min the curve is approaching asymptote but has not as yet come close enough to predict that 0.2 will be the limit. As contrasted with a pool with a single outlet where the limit is the *whole* dose the limit now is an unknown fraction of dose. It is important to predict what it will be because, as shown in Eq. (15), this plateau is the ratio k_{ca}/k_{aa}. Subtraction of points on the observed curve from the true asymptote, as was the case for Model A of Figure 40, will give a simple exponential decay curve. It will be $(k_{ca}/k_{aa})e^{-k_{aa}t}$.

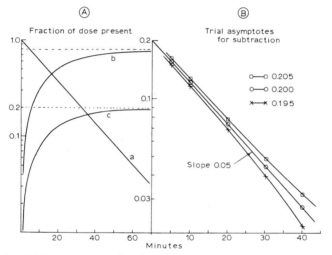

FIG. 41. (A) Time curves for fraction of dose present in labeled pool *a* and collecting reservoirs *b* and *c* of Model B, Figure 40. (B) The plotted curves are fitted to points obtained via subtraction of serial values on curve *c* (part A) from trial asymptotes listed.

(Confirm this by rearranging Eq. (15) and subtracting it from k_{ca}/k_{aa}.) But inspection alone will give a rather poor estimate of the true value of the asymptote. The approach is to choose several trial values and subtract a series of early points on the observed curve from each trial value, then plot the respective curves of subtraction. Illustrative trial values will be 0.195, 0.20, and 0.205. A plot of the separate curves of subtraction from these trial asymptotes is shown in Figure 41B. The choice of 0.205 gives a series of points which follow a slightly concave line on a semilog plot, whereas the choice of 0.195 gives a slightly convex line. Because the line being plotted is a simple exponential function it should be straight. Therefore the correct choice lies somewhere between these two trial values. It, of course, is known in this instance to be 0.2. If it were not known, the error of estimate (between the two curving lines) could not be more than 2%, but of course such a perfect alignment of data points would not be expected in an experimental setting. The line drawn with 0.2 as the chosen value has a measured slope of 0.05 which is k_{aa}, and a zero intercept of 0.2. Its equation may be compared to the theoretical

$$0.2e^{-0.05t} = (k_{ca}/k_{aa})e^{-k_{aa}t}$$

Thus, $k_{ca}/k_{aa} = 0.2$. Substitute 0.05 for k_{aa} and get 0.01 for k_{ca}. Because all of the dose is ultimately in the combined contents of b and c the two asymptotes must add to the entire dose which here is unity because quantity is expressed as fraction of dose

$$k_{ba}/k_{aa} + k_{ca}/k_{aa} = 1$$
$$k_{ba}/0.05 + 0.01/0.05 = 1$$
$$k_{ba} = 0.04$$

The method is applicable for determination of all rate constants regardless of the number of outlets (n), provided $n - 1$ are sampled serially for cumulative output of tracer versus time.

11.   The foregoing approach has been used to study the kinetics of transport of blood iodide to the thyroid in competition with renal removal (Keating et al., 1947). If, in Model B of Figure 40, b is called thyroid and c urinary output, the rate constant to thyroid is calculable from the curve of accumulation in urine. Cumulation curves of splenic uptake of intravenously injected labeled red cells have been used to assess the rapidity of removal of such cells from blood (Holzbach et al., 1964), however the individual rate constant for splenic removal (e.g., k_{ba}) as compared to that for overall removal (k_{aa}) was not calculable because the numerical value of the asymptote in terms of dose was not known. If the geometry of an external radiation detector were such that it measured the true quantity of tracer in spleen, then at asymptote the fraction of dose in spleen would permit a calculation of rate constant to spleen alone. For example, in Model B of Figure 40, consider that reservoir b is spleen and c is " all else." The fraction of dose to b at asymptote in this instance is measured to be 0.8. Thus,

$$0.8 = k_{ba}/0.05$$
$$k_{ba} = 0.04$$

An external detector not measuring the actual fraction of dose but simply a count rate which is proportional thereto will permit calculation only of the overall turnover rate constant of the donor pool. For example, the observed curve in Eq. (17) multiplied by a factor (to represent detection of only a fraction of the tracer in spleen) will not affect the exponential slope, k_{aa}. Subtraction from asymptote still yields a curve with a slope of 0.05.

MULTIPLE INTERCHANGING POOLS

Equations for external accumulation

12. The three-pool system of Figure 40C is the same as that of Figure 20G except that the outlets from pool b and c now go to cumulative reservoirs d and e. Subscripts ob and oc of Model G become respectively db and ec. From Eq. (5) of Chapter 6 with the second point in time being any time, t, rather than ∞, the fraction of dose built up in reservoir d is

$$q_d/q_{a0} = k_{db} \int_0^t (K_1 e^{-g_1 t} + K_2 e^{-g_2 t} + K_3 e^{-g_3 t})\, dt \qquad (19a)$$

Integrated,

$$\frac{q_d}{q_{a0}} = k_{db} \left[\frac{K_1}{g_1}(1 - e^{-g_1 t}) + \frac{K_2}{g_2}(1 - e^{-g_2 t}) + \frac{K_3}{g_3}(1 - e^{-g_3 t}) \right] \qquad (19b)$$

or

$$\frac{q_d}{q_{a0}} = k_{db} \left[-\frac{K_1}{g_1}e^{-g_1 t} - \frac{K_2}{g_2}e^{-g_2 t} - \frac{K_3}{g_3}e^{-g_3 t} \right] + k_{db}\left[\frac{K_1}{g_1} + \frac{K_2}{g_2} + \frac{K_3}{g_3} \right] \qquad (19c)$$

The last term is the asymptote. Substituting an assigned value of 0.03 for k_{db} and using the values previously established for gs and Ks in Figure 20G and Table IIIG,

$$\frac{q_d}{q_{a0}} = 0.03 \left[+\frac{0.885}{0.5}e^{-0.5t} - \frac{0.696}{0.1}e^{-0.1t} - \frac{0.189}{0.01}e^{-0.01t} \right] + (0.03)(24.1) \qquad (20a)$$

$$\frac{q_d}{q_{a0}} = +0.053e^{-0.5t} - 0.209e^{-0.1t} - 0.567e^{-0.01t} + 0.723 \qquad (20b)$$

This is the predicted observed curve for accumulation in reservoir d.

Curve analysis

13. Assume that data points for the foregoing model are recorded for 200 min and that the foregoing equation for the curve is to be derived by curve analysis. As contrasted with the output from a single pool (as in Figure 40B) a *continuous* straight line cannot be obtained by subtraction from trial asymptotes. Nevertheless, subtraction of an asymptote of 0.72 will yield a curve terminating in a straight segment *beyond* 60 *min*. The first peeling, therefore, gives a curve, the terminal segment of which extrapolates as the line $-0.57e^{-0.01t}$. The negative coefficient appears because in peeling the subtraction is actually that of the asymptote from the curve, all points of which lie below the asymptote. The last exponential term and asymptote of Eq. (20b) are now known. If the curve is accurately defined a second peeling will give the second exponential term as $-0.21e^{-0.1t}$. These two intercepts added together give -0.78. Addition of the first unknown intercept to these two must give a total of -0.72 so that, at $t = 0$, when all the exponential terms are unity, the value of the function is zero, as no tracer has as yet appeared outside the labeled pool. Therefore the first intercept must be $+0.06$ (as the closest practical approximation). So much for the intercept-coefficients of desired Eq. (20b). But in a biological system the experimental curve will not be sufficiently accurate to draw a meaningful final subtracted component for the first term of the equation and thereby determine the slope, 0.5. This is because the presence of the slopes in the denominator of the exponential terms in Eq. (20a) reduces the relative contribution of the early components to such an extent that error in dissection at early points in time is greatly magnified. In any event the equation for the curve in itself would not be useful for compartment analysis without further information. If k_{db} were known to be 0.03 this would permit calculation of the K coefficients for the donor pool b (numerators of exponential terms in Eq. (20a)). These are useful in compartment analysis. Equating the previously estimated peeled intercept of 0.57 to its counterpart in Eq. (20a) gives for the last term,

$$0.57 = 0.03K_3/0.01, \qquad K_3 = 0.19$$

Note that via the development of Eq. (19c) from Eq. (19a) the signs of coefficients K are reversed in coefficients for Eq. (19c). Since Eq. (20b) is the numerical counterpart of Eq. (19c), the negative 0.57 requires that $K_3 = +0.19$. Likewise $K_2 = +0.7$, and because the sum of the three K coefficients must be zero, $K_1 = -0.89$. Finally for the first term, with g_1 being the unknown,

$$0.06 = (0.03)(0.89)/g_1, \qquad g_1 = 0.45$$

This compares with the correct value of 0.5 so that the equation for tracer in donor pool b is approximated. But manipulations such as these, while of theoretical interest, will have little use in a biologic system where a chosen model is assigned pools via the device of lumping, and where curves are based on error-prone data points. Accumulation curves as such may have little practical use except to indicate the relative approach to complete recovery of tracer among a series of products which correspond to outlet sites from the system. This deserves further discussion.

An indeterminate system

14. Model D of Figure 40 is indeterminate except that labeled species *a* forms excreted products *b* and *c*. Sampling of pool *a* (e.g., blood) for SA gives irreversible disposal rate of tracee atoms of species *a* (DR_a) via the Stewart–Hamilton equation (Chapter 6). The next step is to determine the rate of conversion of *a* to *b* and to *c*. The first simple assumption is that the fraction of tracer atoms placed in *a* which leave in a given product will be the same as the fraction of unlabeled tracee atoms moving similarly from *a*. Thus if 80% of the dose is ultimately recovered in product *b* this means that 80% of precursor atoms, *a*, are converted to *b*. Then the rate of conversion of *a* to *b* is

$$\text{rate of conversion} \quad a \quad \text{to} \quad b = 0.8 \; DR_a$$

The remaining 20% will be recovered in *c*, and its rate of formation from *a* will be $0.2 \; DR_a$. If the curves of cumulative output of tracer are plotted they will approach respective asymptotes much in the manner of the curves in Figure 41A. In a system such as this where *all output products are identified and collected* the absolute asymptotes of ultimate output need not necessarily be estimated as such. In Figure 41A the fractions of dose in *b* and *c* respectively at 60 min are approximately 0.75 and 0.19. Thus, the fraction of dose to *b* is $0.75/(0.75 + 0.19) = 0.798$, and that to *c* is $0.19/(0.75 + 0.19) = 0.202$ at this point in time. This is close to the ultimate exact distribution between the two. The advantage in constructing curves of accumulation for all products is that judgment may be made as to whether all curves are similarly approaching asymptote. For an example of a system where one lags behind take the model of Figure 29, and for purposes of exaggeration make the size of pool *c* 12 instead of 4. The three curves for accumulation will be

$$\text{From } a = -0.16e^{-0.5t} - 0.2e^{-0.1t} + 0.36$$
$$\text{From } b = +0.052e^{-0.5t} - 0.26e^{-0.1t} + 0.208$$
$$\text{From } c = -0.004e^{-0.5t} - 0.027e^{-0.1t} + 0.463e^{-0.01t} + 0.432$$

A plot of these curves will show that those from *a* and *b* rise sharply, and at 40 min are near 0.36 and 0.20, respectively, i.e., visually quite close to asymptote. On the other hand the curve from *c* is obviously still rising and is only at 0.12 as compared to its ultimate predicted limit of 0.43. Its 40-min value cannot be used, although those for *a* and *b* are seen by the shape of the curves to be directly usable. Conversion rate of *a* to *b*, for example, is approximately: $(0.28)(0.2) = 0.056$. That to *c* can be approximated only by the difference between total disposal rate and that via $a + b$.

Sink Effect

A THREE-POOL EXAMPLE

15. Pool c in Figure 42 represents a *very large* compartment in comparison with the other two pools in an open interchanging system. Tracer in pool b can move either irreversibly out of the system to the accumulating reservoir d, or reversibly into the large pool c. The pertinent effect of this large pool is to sequester tracer in the early period of time shown on the graph as though it were irreversibly lost from pools a and b. Note that the shape of the curve for quantity of tracer in pool c resembles that for cumulative loss from the system into reservoir d. Tracer entering pool c is, in a sense, lost because of the sink effect of such a relatively large compartment. Although rate of inflow to pool c (0.023 mg/min) equals rate of return flow, the fractional rate constant for return flow to b from c is only $0.023/265 = 0.000087$ as compared to $0.023/1.5 = 0.015$ for input transfer to c from b. Of course the curve for pool c after about 150 min is not actually flat. It is falling very slightly just as the curve for output to d is rising very slightly toward 1. The extent of this drift can be determined from the equations for the curves,

$$q_a/q_{a0} = 0.6e^{-0.4t} + 0.3999e^{-0.05t} + 0.0001e^{-0.00007t}$$
$$q_b/q_{a0} = -0.7428e^{-0.4t} + 0.7426e^{-0.05t} + 0.0002e^{-0.00007t}$$
$$q_c/q_{a0} = 0.029e^{-0.4t} - 0.23e^{-0.05t} + 0.20e^{-0.00007t}$$
$$q_d/q_{a0} = 1 - (-0.114e^{-0.4t} + 0.913e^{-0.05t} + 0.200e^{-0.00007t})$$

The extremely small constant (0.00007) in the last exponent is reflected in the relatively flat portion of the curves. This quasiplateau for pool c approximates a true plateau of irreversible loss.

16. Assuming that tracer flowing to pool c is, for practical purposes, irreversibly lost from pools a and b during the first two hours (as is that to reservoir d), the plotted curves for a, c, and d of Figure 42 will permit a reasonably satisfactory calculation of several rates. The foregoing formulas are not pertinent for this purpose. Assume that serial observations are made for the SA of pool a, and for the cumulative amount of tracer in c and d during the first 100 min. The ultimate flattening of the curve for pool a is disregarded. Its straight portion from 10–100 min is extrapolated to infinity and to zero time to give the following equation via peeling,

$$SA_a/SA_{a0} = 0.6e^{-0.38t} + 0.4e^{-0.05t}$$

Dose (1) divided by the integral of this expression gives 0.116 for "irreversible" disposal rate of species a. At 100 min the curves for pool c and collecting reservoir d are seen to be near effective plateau values of 0.2 and 0.79, respectively. Thus, the rate of conversion of species *a* to species *c* is (0.116) (0.2) = 0.023, and that to species *d* is (0.116)(0.79) = 0.092. These estimates are very good approximations of rates to *c* and *d* as shown. Keep in mind that the rate of return movement of material to *b* from *c* is not given by this calculation. It may or may not equal the rate of flow into *c*. This method of

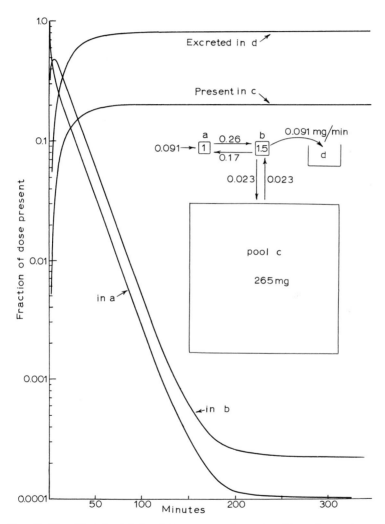

Fig. 42. The effect of a sink (pool *c*) as compared to external loss of tracer (reservoir *d*). Rates accompanying arrows are in units of milligrams per minute. Numbers in boxes are pool size in milligrams.

estimating rate of disposal of material in a primary labeled pool has been used to study the conversion of glucose carbon to CO_2, fat, protein, glycogen, and urinary material (Shipley *et al.*, 1967).

DEFINITION OF DISPOSAL RATE

17. In the foregoing example disposal rate was calculated to be 0.116, yet, note that inflow rate, F_{ao}, in the model is only 0.091, which matches output from the entire system to external collecting chamber as species *d*. This emphasizes the importance of defining disposal rate. In this instance, it includes rate to the sink *c*. Kinetic events which dominate the observed curves in the first 100 min have been dissected from events transpiring thereafter. The ultimate disposal rate to the outside of the entire system is only 0.091. For example if species *a* is glucose carbon, species *c* is carbon of fat, and species *d* is excreted CO_2 carbon, the glucose carbon going to fat is irreversibly disposed only in a relative sense. The *in vivo* model is obviously more complex than the one shown, but the sink principle applies to those compounds which represent a sizeable portion of body mass with relatively small turnover constants.

CHAPTER 9

CONSTANT INFUSION OF TRACER

Estimation of Rate without Compartment Analysis

1. As a method of introducing tracer, sustained delivery over a protracted period of time antedates abrupt delivery as a single dose. Rittenberg and Schoenheimer (1937) fed deuterium-enriched water over a period of many days and plotted the time curve for build-up of incorporated ^2H in fat. Continuous delivery of tracer via an intravenous infusion over a period of hours, first used by Stetten *et al.* (1951) and popularized by Steele *et al.* (1956), has been employed to study the rate of production and disposal of glucose and a variety of other compounds as well.

TURNOVER RATE OF A SYSTEM

Single pool

2. Turnover rate is input–output rate of unlabeled material in a steady-state system. The principle by which tracer, infused constantly in steady amount, measures this rate is easily grasped without complex mathematics by examining the single pool of Figure 43A. Endogenous input rate, F_1, of 5 mg/min is matched by outflow rate, F_2. Carrier-free tracer matching the

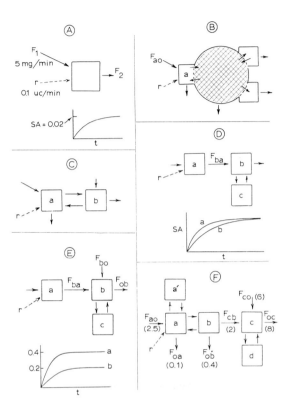

FIG. 43. Models to accompany calculations when tracer is given by constant infusion at rate, r. See text.

species in the pool is constantly infused at any rate, r. In this example r will be 0.1 μCi/min. Serial measurements of specific activity of the constantly mixed material in the pool will yield a rising curve which approaches a flat horizontal asymptote at 0.02 μCi/mg (see graph of Figure 43A). The rising segment of the curve represents progressive washout of preexisting unlabeled material accompanied by simultaneous replacement by a steadily supplied mixture of infused tracer plus an endogenous input in the *constant proportion of* 0.1 μCi *to* 5 *mg*. In other words the SA of this joint input is 0.02 μCi/mg. Obviously, in course of time the content of the pool will consist totally of material having an SA of 0.02. Call this plateau value SA_E for SA at equilibrium. Then, as just explained,

$$SA_E = r/F_1 \qquad \text{or} \qquad F_1 = r/SA_E$$

In this instance $F_1 = 0.1/0.02 = 5$ mg/min. This also is output rate, F_2, i.e., it is turnover rate for this simple system in steady state.

Complex models

3. Model B of Figure 43 differs from Model A in that any number of reversibly connected secondary pools may exist. The configuration is largely unknown (as represented by the cross-hatched region) except for the fact that *all externally derived input* to the system *is via the infused pool* (pool *a*). This is to say that species *a* is formed anew only *directly* via F_{ao}. Given sufficient time and all parts of the system (e.g., secondary compounds) will attain the same constant and identical SA at equilibrium. Thus, an equation may be written to calculate overall input–output rate for the system as a whole by sampling any pool or site in such a system,

$$\text{rate of overall input–output} = r/SA_{iE} \qquad (1)$$

The symbol SA_{iE} signifies the SA value at equilibrium at any site, *i*, in the system. Such flexible option in site for sampling does not hold for Model C because part of the endogeneous entry of unlabeled material is into a region (reversibly convertible compound) other than to species *a*. Entrance occurs also at *b*. If pools *a* and *b* are not interchanging so rapidly that they behave effectively as a single joint pool the SA of noninfused pool *b* will remain continually lower than that of infused pool *a*. Total turnover rate of the system is not given by sampling either pool. Models which qualify for assessing *overall* turnover rate by sampling a given pool are the same as those which qualify for application of the Stewart–Hamilton equation (Chapter 5, paragraph 19).

PRODUCTION–DISPOSAL RATE

4. Net production rate or its equivalent rate of irreversible disposal in a steady state system is calculable for tracee atoms of the *species of the infused pool* by observing the plateau SA of the *infused pool* (not necessarily other pools) regardless of the configuration of the system, or whether other sites of input and output exist. Note that this also was true for application of the Stewart–Hamilton equation after a single dose of tracer to the sampled pool. In Model C the production rate of tracee atoms in species *a*, i.e., PR_a, is $F_{ao} + F_{bo}[F_{ab}/(F_{ab} + F_{ob})]$. This is the portion of input to the system which, when reaching pool *a*, is *new* to pool *a*. It is not *total turnover rate* of species *a* atoms which would be $F_{ba} + F_{oa}$ or $F_{ab} + F_{ao}$; however it is turnover rate of *a* atoms in Model D where no recycling occurs. The derivation of the equation for production–disposal rate of the tracee in infused pool *a* is as follows: *After plateau is achieved* the amount of tracer leaving pool *a* irreversibly in a given period of time (T) will equal the amount entering for the first time. The latter is infusion rate multiplied by the time span. Thus, the amount entering is $r \cdot T$. This should be equated to the amount of tracer accompanying the portion of unlabeled material in pool *a* which is irreversibly lost in the same time. The amount of unlabeled material lost in time span, T, is the product of disposal rate and the time span, i.e., it is $DR_a \cdot T$. Further multiplication by the SA of species a atoms at plateau (SA_{aE}) gives amount of tracer lost during T. Thus, equating the amount of tracer entering with the amount leaving,

$$r \cdot T = DR_a \cdot SA_{aE} \cdot T$$

Expressing this in terms of DR_a

$$DR_a = (r \cdot T)/(SA_{aE} \cdot T)$$

Of interest at this point is the similarity between this intermediate equation and the Stewart–Hamilton equation for a single dose of tracer. Both are derived under the simple assumption that the amount of tracer administered equals the amount lost. In the foregoing equation for DR_a the numerator is dose delivered (comparable to D in the Stewart–Hamilton equation) and the denominator is the area under the SA curve during which an amount of tracer equal to the dose is lost. (See Chapter 5, paragraph 5, and Chapter 6, paragraph 1.) Before cancelling T in numerator and denominator the equation is actually D/A. Afterward, it becomes

$$DR_a = r/SA_{aE} \tag{2}$$

Pool a is the infused pool in any unrestricted system. The right-hand side of this equation is the same as that of Eq. (1) except that in an unrestricted system it pertains strictly to the SA of the traced atoms in the *infused* pool. For disposal rate of tracee from a secondary pool, a correction must be made for fraction of tracer reaching the pool, e.g., for SA of pool b when infusion is to a:

$$DR_b = \frac{rF_{ba}/(F_{ba} + F_{oa})}{SA_{bE}} \tag{2a}$$

Note that in steady state Eqs. (2) and (2a) also apply respectively to net production rates, PR_a and PR_b.

FRACTIONAL CONTRIBUTION OF NONREVERSIBLE POOL

5. In Models D and E of Figure 43 the pertinent feature for consideration is the irreversible transfer to b from a (F_{ba}). In Model D the plateau SA for pool b, although attained much later, is identical to that for pool a. But in model E where an independent source of input (F_{bo}) enters pool b from the outside, sampling of pools a and b will give the two curves as shown in the graph. Query: What are the rates F_{ba} and F_{bo}? With an infusion rate of $1 \mu Ci/$ min, and plateau SA for pool a of $0.4 \mu Ci/mg$, Eq. (2) gives 2.5 mg/min for F_{ba}. The plateau SA for pool b is $0.2 \mu Ci/mg$. Independent input to b via F_{bo} cuts the SA to half that of pool a. Consequently, F_{bo} must equal F_{ba}, and each contributes half of total input to b (and half of total output, F_{ob}). Thus,

$F_{bo} = F_{ba} = 2.5$ mg/min, and $F_{ob} = 5$ mg/min. The basic formula for fraction of input to b coming from a (or fraction of external discharge from b coming from a) is

$$F_{ba}/(F_{ba} + F_{bo}) = SA_{bE}/SA_{aE} = F_{ba}/F_{ob}$$

Specifically in this instance,

$$\text{fraction of overall input to} \quad b \quad \text{from} \quad a = 0.2/0.4 = 0.5$$

Thus 50 % is supplied from a. For complete solution,

$$2.5/(2.5 + F_{bo}) = 0.2/0.4 = 2.5/F_{ob}$$

Simple algebra yields both F_{bo} and F_{ob}. Model F introduces complexity. Tentatively, it might be considered a partial model for transfer of glucose carbon from blood glucose in pool a. F_{ao} would be rate of formation of new glucose carbon from gut and liver; a', that in interstitial fluid; F_{oa}, rate of possible urinary excretion; b, carbon of nonglucose intermediates being fed via F_{ob} to fats, etc.; c, bicarbonate carbon of blood; F_{co}, non glucose source of the latter; F_{oc}, excretion of CO_2 carbon in breath. Again assume that $r = 1.0$, and that at plateau of 0.4 for glucose carbon this same SA is established for a' and b, whereas SA of c and its side pool d (e.g., cell bicarbonate) is established at 0.1.

$$F_{ao} = r/SA_{aE} = 1/0.4 = 2.5$$
$$F_{oc} = 8 \quad \text{(measured chemically as expired } CO_2 \text{ carbon)}$$
$$F_{cb}/F_{oc} = SA_{cE}/SA_{aE} = 0.1/0.4 = 0.25$$

This would mean that 25 % of CO_2 carbon is from glucose carbon. Then

$$F_{cb} = (8)(0.25) = 2$$
$$F_{co} = F_{oc} - F_{cb} = 8 - 2 = 6$$
$$F_{oa} + F_{ob} = F_{ao} - F_{cb} = 2.5 - 2 = 0.5$$
$$F_{oa} = 0.1 \quad \text{(measured chemically as urinary excretion)}$$
$$F_{ob} = 0.5 - 0.1 = 0.4$$

Although this appears straightforward, certain features of the actually more complex physiologic system introduce practical problems. For example, the portion of F_{ao} from liver is not truly an *external* source of carbon nor is F_{ob} a definitive exit. Some tracer carbon moving via the latter reaches liver, e.g., as lactate, and appears in liver-derived glucose. If tracer is equilibrated in this circuit the rate of donation of recycled carbon is excluded from the estimate of total F_{ao}. What is given is net production rate of glucose carbon, i.e., carbon *new* to blood glucose. This is PR_a (paragraph 4). And in the rat, at least, the equilibration with potentially recycling intermediates is very prompt (Shipley, 1967). On the other hand, protein-derived carbon, being from a sink, and therefore effectively nonlabeled, is included in the calculated value. Note, however, that each of the aforementioned plateau values is actually only a quasi-plateau which hopefully will be recognizable before tracer returns in significant amount from compounds considered sinks. But eventually the tracer will equilibrate even in all sinks so that carbon leaving therefrom will have the same SA as that entering. In a fasting animal having no truly external source of glucose carbon the theoretical *ultimate* limit for a uniform SA of *all* communicating body compartments is that of the *infused* glucose itself. Calculated rate of production of "cold" glucose carbon then would simply be that for infusion rate of carrier atoms accompanying tracer, and zero for endogenous production. If, by constant infusion, one intends to measure total *turnover* rate of the glucose *molecule* this will be possible by choosing a component, which unlike carbon, does not recycle, i.e., it is promptly and irreversibly lost after leaving blood glucose. Irreversible disposal and turnover rates then are the same. Tritium label for nonrecycling hydrogen at position 2 has been used for this purpose (Hetenyi and Mak, 1970). Although a complete solution for transfer rates is not possible in an unrestricted reversible model (e.g., Figures 33B and 43C), the fraction of any secondary species, *i*, derived from labeled *a* is given by the ratio SA_{iE}/SA_{aE}. Proof is similar to that for a single dose (Chapter 6, paragraph 3) except that respective products, $(SA_{iE})(T)$ and $(SA_{aE})(T)$, replace the integrals.

DOUBLE LABELING

A two-pool reversible system

6.   The approach is similar to that for a single dose as described in Chapter 6, paragraphs 10 and 11. Equations (13) and (14) of that chapter are modified as follows to accomodate constant infusion as shown in model A of Figure 44,

$$PR_a = \frac{r^a}{SA_{aE}^a} = \frac{r^b[F_{ab}/(F_{ab} + F_{ob})]}{SA_{aE}^b} = F_{ao} + F_{bo}\frac{F_{ab}}{F_{ab} + F_{ob}} \qquad (3)$$

$$PR_b = \frac{r^b}{SA_{bE}^b} = \frac{r^a[F_{ba}/(F_{ba} + F_{oa})]}{SA_{bE}^a} = F_{bo} + F_{ao}\frac{F_{ba}}{F_{ba} + F_{oa}} \qquad (4)$$

PR_a and PR_b are production rates of species *a* and species *b*, respectively, *r* is rate of infusion of tracer, and the superscript to *r* indicates which species is infused. Likewise, for SA the superscript denotes which species was infused, while the subscript is the species sampled for plateau SA. Solution for all rates proceeds as described in Chapter 6 (including Eqs. (15) and (16) of that chapter). The hatched region between the two pools in Figure 44A indicates the possible presence of intermediary species.

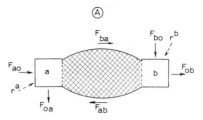

FIG. 44. More models to illustrate constant infusion. See text.

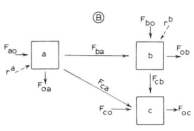

A three-pool nonreversible system

7. A model such as that of Figure 44B has been used by Bergman *et al.* (1968) to determine rates of glycerol conversion (pool *a*) to glucose (pool *b*) and CO_2 (pool *c*). In this one-way model production rate of atoms of species *a* is the same as its turnover rate, and fractional contributions are calculable from ratios of specific activities as in paragraph 5. In the following set of equalities column (a) is total input rate to pools *a, b, c,* and again *c* successively downward, and column (d) is corresponding total output rate. Column (b) is plateau SA in relation to direct infusion into pools *a* and *b*, respectively. Column (c) is respective plateau SA of pool *b*, then *c*, and again *c*, in relation to fraction of infusion to pool *a* reaching these respective pools (Eqs. (6) and (7)), followed by plateau SA of pool *c* in relation to fraction of infusion to pool *b* reaching pool *c* (Eq. (8)),

$$\text{(a)} \qquad \text{(b)} \qquad \text{(c)} \qquad \text{(d)}$$

$$F_{ao} = \frac{r^a}{SA^a_{aE}} \qquad\qquad = F_{ba} + F_{ca} + F_{oa} \qquad (5)$$

$$F_{bo} + F_{ba} = \frac{r^b}{SA^b_{bE}} = \frac{F_{ba}}{F_{ao}}\frac{r^a}{SA^a_{bE}} \qquad = F_{ob} + F_{cb} \qquad (6)$$

$$F_{co} + F_{ca} + F_{cb} = \frac{F_{ca}}{F_{ao}}\frac{r^a}{SA^a_{cE}} \qquad = F_{oc} \qquad (7)$$

$$F_{co} + F_{ca} + F_{cb} = \frac{F_{cb}}{(F_{bo} + F_{ba})}\frac{r^b}{SA^b_{cE}} = F_{oc} \qquad (8)$$

The equalities (a5) and (b5) give F_{ao}, which when substituted in (c6) and equated to (b6) gives F_{ba}. F_{bo} follows from (a6) and (b6), and F_{ca} comes from (c7) along with (d7), which is measured experimentally. Then, F_{cb} follows from (c8), (d8), and (a6). F_{oa} emerges from (a5) and (d5), and F_{ob} from (d6) when values already calculated are substituted.

PRIMING DOSE

8.   The delay in reaching asymptote can be shortened by injecting a priming dose of tracer before beginning the infusion. Figure 45 shows a series of SA curves with an assortment of priming doses, P, placed in a system consisting

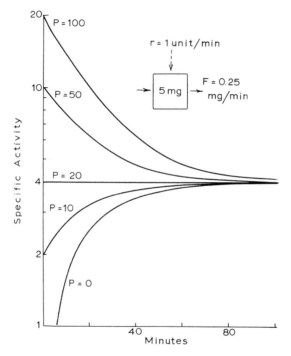

FIG. 45. Effect of size of priming dose, P, on the SA curve of a single pool system receiving constant infusion of tracer at a fixed rate ($r = 1$).

of the *single pool* as shown. Infusion of tracer proceeds at a rate of 1 unit/min in all instances. The lowermost curve represents infusion with no priming dose. The second from bottom, representing 10 units of primer rises to plateau more promptly. With 20 units the plateau is achieved instantly. Doses larger than 20 times infusion rate yield curves which fall toward asymptote instead of rising. Note that the large dose of 100 units actually postpones achievement of the horizontal phase. The optimum dose for near-instant plateau would be predictable in a biologic system if it indeed consisted of a *single* pool of known size having known input–output rate. But if these are already known the experiment is pointless. The proper ratio of priming dose to infusion rate for earliest attainment of plateau will be found by trial and error. Whether the investigated moiety of a biologic system can be treated as a lumped equivalent of a single pool will depend on the system. This will be discussed under three-compartment systems.

Compartment Analysis

SINGLE POOL

9.   Until this point the constant infusion method has been aimed at measurement of rates of input–output or interchange. If, in addition to observing SA at plateau, the more complete SA versus time function is obtained by serial sampling beginning at or near zero time, it is theoretically possible to calculate rate constants and pool sizes in a manner similar to the approach for a single injection as described in Chapters 1–4. Practical limitations in applying the method are just as real as was true for the single injection method. In other words a "blind approach" is not feasible. The configuration of the model must be known and the magnitude of possible errors arising from injudicious lumping of two or more pools as though they were one must be fully appreciated.

No priming dose used

10.   For the single pool of Figure 45 the differential equation for rate of change of quantity of tracer (q) in the single pool with constant infusion at rate r is

$$dq/dt = r - kq \qquad (9)$$

where k is the rate constant of efflux. Integration provides the equation for q, as a function of time,

$$q = (r/k)(1 - e^{-kt})$$

For specific activity, division by pool size (Q) gives

$$q/Q = \text{SA} = (r/F)(1 - e^{-kt}) \qquad (10)$$

At plateau (i.e., $t = \infty$) the exponential term in Eq. (10) is equal to zero, and $SA_E = r/F$. Extrapolate this plateau value ($SA_E = 4$ in Figure 45) to zero time and serially subtract values on the lowermost curve from it. Algebraically, r/F minus Eq. (10) gives a new function to be designated $f(t)$,

$$f(t) = (r/F)e^{-kt}$$

This is the equation for the derived curve to be plotted after subtraction. It is a simple exponential function with slope k. In the present example it will be

$$f(t) = 4e^{-0.05t}$$

Thus $k = 0.05$. From Eq. (1), F already was known to be 0.25. Then $Q = F/k = 0.25/0.05 = 5$ mg. If the plot of the curve of subtraction is not a straight line on semilog coordinates this means either that the system effectively behaves as a multicompartment system or that the true asymptote has been misjudged. A method for judging the latter in a single pool system is given in Chapter 8, paragraph 10.

Priming dose used

11.   When a priming dose of tracer (P) is introduced at zero time and constant infusion at rate r is continued thereafter the differential equation for rate of change of quantity of tracer is the same as Eq. (9) but integration gives a different result because at zero time q is P rather than zero. The expression for quantity of tracer as a function of time becomes

$$q = \frac{r}{k} + \left(P - \frac{r}{k}\right)e^{-kt} \tag{11}$$

For specific activity, divide both sides by Q,

$$\frac{q}{Q} = SA = \frac{r}{F} + \left(\frac{P}{Q} - \frac{r}{F}\right)e^{-kt} \tag{12}$$

Thus, for any curve in Figure 45, Eq. (12) becomes

$$SA = 4 + (P/Q - 4)e^{-kt}$$

Again, at asymptote, $SA_E = r/F$, and $F = \frac{1}{4} = 0.25$. If, for any curve except the middle one, serial points are subtracted from the asymptotic value 4, the theoretical curve will be

$$f(t) = (4 - P/Q)e^{-kt}$$

This also is a simple exponential. If priming dose is 10, the equation for the observed subtracted curve will emerge as

$$f(t) = 2e^{-0.05t}$$

The rate constant, again, is equivalent to the slope of the line. Equate the coefficient (intercept) of the subtracted curve with the theoretical,

$$2 = 4 - (10/Q)$$
$$Q = 5$$

or with a priming dose of 100, subtraction will give

$$f(t) = -16e^{-0.05t}$$

Again, equating coefficients,

$$-16 = 4 - (100/Q)$$
$$Q = 5$$

If the priming dose is 20, no subtraction is necessary. The flat horizontal line reaching from 0 to ∞ means that $P/Q = r/F$. The former is the instantaneous SA at t_0, and the latter is ultimate SA. Being equal, the second term on the right in Eq. (12) becomes zero. Then $r/F = P/Q = 20/Q = 4$ and $Q = 5$.

THREE-POOL INTERCHANGING SYSTEM, EQUATIONS

No priming dose

12. The model is unspecified (e.g., Models B–I, Figure 20). Infusion is to pool a. The equation for quantity of tracer in pool a is derivable as follows: Consider that infusion during successive short intervals Δt supplies a sequence of small separate bits of tracer, $q'_{a0} + q''_{a0} + \cdots$, etc. For the first separate bit the amount of tracer in pool a as a function of time is

$$q_a' = q'_{a0}(H_1 e^{-g_1 t} + H_2 e^{-g_2 t} + H_3 e^{-g_3 t})$$

(See Chapters 2 and 3 and Appendix II.) The bit of dose, q'_{a0} introduced to a during Δt also may be expressed as infusion rate times the interval Δt. Thus,

$$q_a' = r \cdot \Delta t \, (H_1 e^{-g_1 t} + H_2 e^{-g_2 t} + H_3 e^{-g_3 t})$$

Similar expressions apply to $a''_{a0}, \ldots$, etc. The sum of all such bits is that for all tracer in pool a at time t. Call the complex exponential in parentheses $f(t)$,

$$q_a(t) = \sum r \cdot f(t) \cdot \Delta t$$

In the integral form

$$q_a = r \int_0^t f(t) \, dt$$

Also,

$$SA_a = (r/Q_a) \int_0^t f(t) \, dt$$

Integrate the latter,

$$SA_a = \frac{r}{Q_a}\left[\frac{H_1}{g_1}(1 - e^{-g_1 t}) + \frac{H_2}{g_2}(1 - e^{-g_2 t}) + \frac{H_3}{g_3}(1 - e^{-g_3 t})\right]$$

or

$$SA_a = \frac{r}{Q_a}\left(-\frac{H_1}{g_1}e^{-g_1 t} - \frac{H_2}{g_2}e^{-g_2 t} - \frac{H_3}{g_3}e^{-g_3 t}\right) + \frac{r}{Q_a}\left(\frac{H_1}{g_1} + \frac{H_2}{g_2} + \frac{H_3}{g_3}\right) \quad (13)$$

For quantity of tracer (q_a) rather than SA_a, omit Q_a. The constant term on the right is the asymptote (plateau value). Equation (13) may be modified to yield an expression for any pool. For example, in pool b in any of the afore-mentioned three-pool systems use coefficients K in place of H, and Q_b in place of Q_a. Comparable changes are made for pool c. For definitions of H, K, and L, see Chapters 2 and 3, and Appendix II. They are normalized coefficients applying to curves for a single dose of tracer. With an infusion rate, r, of 2 units/min, and $Q_a = 1$, the following is the curve for pool a obtained by substituting numerical values for H and g in Eq. (13) (Eq. (1b), Chapter 3, with dose of 1 rather than 100),

$$SA_a = \frac{2}{1}\left(-\frac{0.7}{0.5}e^{-0.5t} - \frac{0.2}{0.1}e^{-0.1t} - \frac{0.1}{0.01}e^{-0.01t}\right)$$

$$+ \frac{2}{1}\left(\frac{0.7}{0.5} + \frac{0.2}{0.1} + \frac{0.1}{0.01}\right)$$

$$SA_a = -2.8e^{-0.5t} - 4.0e^{-0.1t} - 20e^{-0.01t} + 26.8 \quad (14)$$

As noted in paragraph 4, production–disposal rate for species a is $PR_a = r/SA_E = 2/26.8 = 0.0746$ mg/min in Models B–I, Figure 20. This also is rate of input–output for the system as a whole in all models save H (which has some input–output avoiding complete transit via a). Compartment analysis for assessment of rates of interflow and the size of each pool is *theoretically* possible. The observed curve (Eq. (14)) would be the lowermost one in Figure 46. Its asymptote could be predicted (beyond 350 min) as described in Chapter 8, paragraph 13. Subtraction from this asymptote of 26.8 gives

$$f(t) = 2.8e^{-0.5t} + 4.0e^{-0.1t} + 20e^{-0.01t} \quad (15)$$

If this subtraction is plotted and analyzed, the slopes theoretically should emerge, however the slope g_1 (i.e, 0.5), for the same reasons given in paragraph 13 of Chapter 8, cannot be accurately estimated. The following alternate approach to analysis may be possible. The third exponential term of Eq. (15) will emerge in graphic analysis to give a slope of 0.01 and an intercept of 20. The latter is the coefficient -20 in Eq. (14). From the parent equation for

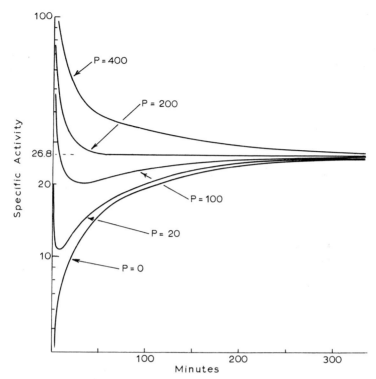

FIG. 46. Effect of size of priming dose, P, on the SA curve for a three-pool interchanging system. Curves are from the primed and infused pool.

Eq. (14), and Eq. (13), the value of 20 is seen to have come from the multiple of the common coefficient r/Q_a and H_3/g_3. Thus,

$$20 = (2/Q_a)(H_3/0.01)$$

Likewise, for the second coefficient (g_2 having emerged in peeling of the subtracted curve given in Eq. (15)),

$$4.0 = (2/Q_a)(H_2/0.1)$$

For the first coefficient of Eq. (15) the slope g_1 cannot be accurately estimated. Thus, at this point, it is one of the unknowns,

$$2.8 = (2/Q_a)(H_1/g_1)$$

From the properties of the H coefficients,

$$H_1 + H_2 + H_3 = 1$$

These comprise four simultaneous equations with five unknowns. If the size of pool a (Q_a) is independently known, the number of unknowns is reduced to four, and solution is possible from the values for H and g (Chapter 3). However, when all is said and done, compartment analysis can be performed more simply and accurately, and without necessarily knowing Q_a, if the single dose technique is used rather than constant infusion.

With priming dose

13. The curves in Figure 46 include four with different sized priming doses. Infusion rate is 2 units/min in all instances. Note that as priming dose is increased to a certain point the plateau is approached more promptly. With 20 and 100 units a double inflection appears. With 200 units this disappears and the simple decline is followed by the earliest attainable plateau. A larger dose delays establishment of plateau. The plateau always will be 26.8, and from Eq. (2), $2/26.8 = 0.0746$, the production–disposal rate of a. The equations for any of the primed infusion curves in Figure 46 may be predicted by combining the equation for a single dose with Eq. (13) for constant infusion. For a single dose of size P to pool a,

$$SA_a = (P/Q_a)(H_1 e^{-g_1 t} + H_2 e^{-g_2 t} + H_3 e^{-g_3 t})$$

Add this to Eq. (13), rearrange, and get

$$SA_a = \left(P - \frac{r}{g_1}\right)\frac{H_1}{Q_a}e^{-g_1 t} + \left(P \div \frac{r}{g_2}\right)\frac{H_2}{Q_a}e^{-g_2 t}$$
$$+ \left(P - \frac{r}{g_3}\right)\frac{H_3}{Q_a}e^{-g_3 t} + \frac{r}{Q_a}\left(\frac{H_1}{g_1} + \frac{H_2}{g_2} + \frac{H_3}{g_3}\right)$$

Again, the constant (last) term is the asymptote (plateau). To construct the curves of Figure 46 via this equation, the set of values for Hs and gs were taken from Chapter 3, paragraph 2. $Q_a = 1$. The equations for the SA curves as shown are

$$SA_a = 277e^{-0.5t} + 76e^{-0.1t} + 20e^{-0.01t} + 26.8, \qquad P = 400$$
$$SA_a = 137e^{-0.5t} + 36e^{-0.1t} + 0 + 26.8, \qquad P = 200$$
$$SA_a = 67.2e^{-0.5t} + 16e^{-0.1t} - 10e^{-0.01t} + 26.8, \qquad P = 100$$
$$SA_a = 11.2e^{-0.5t} + 0 - 18e^{-0.01t} + 26.8, \qquad P = 20$$
$$SA_a = -2.8e^{-0.5t} - 4.0e^{-0.1t} - 20e^{-0.01t} + 26.8, \qquad P = 0$$

One feature of this series is that an exponential term disappears with certain values of P. With the other sized priming doses compartment analysis is possible by the subtraction maneuver described in paragraph 12. The accuracy in defining the first slope, g_1, is substantially enhanced as priming dose is increased. The first exponential term becomes dominant at early time because the preponderant effect then is that of a single dose. But again, if compartment analysis is the goal a single dose with no constant infusion is preferable. The same may be said for determination of production rate of species *a* (which requires nothing more than subtended area of the SA function).

PREDICTION OF ASYMPTOTE

14. In the model of Figure 47 the ratio of asymptote for the SA curve of pool *b* to that for the comparable curve of pool *a* will give the ratio of F_{ba} to F_{ob} (paragraph 5). But with or without primer the curve for pool *b* is still rising at 140 min. It has reached only 5.7, although in this known system it is predicted to reach 6.67. By 300–400 min, it will be within 97–99% of

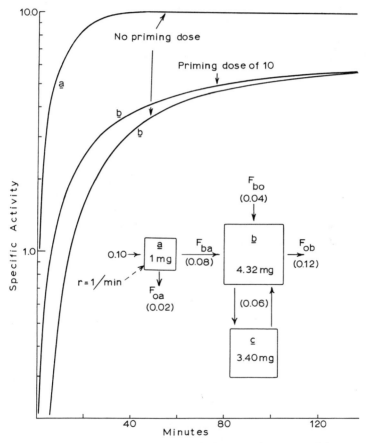

FIG. 47. Model to illustrate prediction of plateau value expected in a secondary pool (*b*).

maximum, but an experimental period of this duration may be impractical. Furthermore, such prolongation might be undesirable in a bona fide biologic system. Perhaps the output F_{oa} is to a sink rather than to the outside (Chapter 8, paragraphs 15 and 16). If it so happened that input F_{bo} were fed by this sink the SA curve for pool b would continue to rise slowly for a very long time and its limit would not pertain to the model as drawn. If the curve is to become reasonably flat in reasonable time its limit may be predicted by the graphic method of Chapter 8, paragraph 13. Steele *et al.* (1959) have used another approach for a glucose–CO_2 system. Consider pool b an end-pool similar to HCO_3. Because pool a yields a curve which promptly levels off, the retarded approach to asymptote in b is ascribable to pool b itself plus its reversible associates (b and c in this instance). Upon labeling b with a single dose of its own species in a separate experiment the decay curve for sampled b will be amenable to graphic analysis. Its equation then can be converted to one which corresponds to constant infusion to b and the nearness to asymptote can be calculated for any point in time. In the system of Figure 47 the SA curve for a single unit dose to b is predicted to be

$$SA_b = 0.186e^{-0.05t} + 0.046e^{-0.01t}$$

If pool b were infused at a rate of 1 unit/min this equation would be altered as predicted by Eq. (13) for a labeled pool,

$$SA_b = -3.71e^{-0.05t} - 4.63e^{-0.01t} + 8.33$$

At $t = 140$ min, the foregoing equation yields a value of 7.2. This divided by the predicted asymptote of 8.33 gives a ratio of 0.86 for fraction of asymptote value attained at 140 min. The observed curve when pool a was infused had reached 5.7 at this same time. Its correction in terms of predicted asymptote is $5.7/0.86 = 6.6$. This compares to the theoretically correct value of 6.7 as given by the equation for the b curve with constant infusion to a (lower curve, Figure 47). This equation, incidentally, is

$$SA_b = +3.37e^{-0.1t} - 5.93e^{-0.05t} - 4.12e^{-0.01t} + 6.67$$

The prediction is not absolutely exact because exclusion of pool a in the trial labeling of b does to some extent affect the late portion of the curve even though a preliminary passage through a causes relatively little holdup (in the present model).

SIZE OF THE SYSTEM

15. Published results with the constant infusion method have included calculations of the joint size of the infused pool and associated interchanging pools. As was emphasized in Chapter 1 for a single injection of tracer (paragraphs 14–23), this may be attended by a large error unless interchange between components of the complex is rapid enough to justify lumping all into the equivalent of a single pool. Three clues assist in judging whether such lumping is a reasonable assumption. Subtraction of the curve of constant infusion, with or without primer, from a well-established asymptote should, on a semilog scale, plot as a straight line (simple exponential function) which remains straight close to zero time. Inflections prior to straightening should be short lived. Also, a certain size of priming dose can be found which will give a very early plateau for the curve of infusion. For example, in Figure 45 the 20 unit primer is perfect. If such perfection is approached the SA near zero time reflects the near-instant dilution of primer by the entire complex, and therefore P/Q = SA of the near-immediate plateau (SA_E). Then $Q = P/SA_E$. A clue to the presence of slowly equilibrating pools is seen in the plotted curves of Figure 46. The doubly inflected contour after priming doses of 20 and 100 is not consistent with a system wherein interchange is relatively rapid. Theoretically, the closest approximation of overall size of this system would be given by a priming dose of 200 because the known equations predict that it gives earliest plateau. Then, calling 26.8 a "near-instant" plateau (which it is not), $Q = 200/26.8 = 7.5$ mg. The correct value will actually depend on the model. For Models B–I of Figure 20 (save H, where all input is not to the labeled pool) the range of overall size (from Table III) is 6.1 to 8.4. Although 7.5 is not an unreasonably poor approximation, an onerous chore in an actual experiment would be to show by a series of trials that 200 is actually the best dose for achieving earliest plateau.

Clearance

16. Clearance is defined in some detail in Chapters 1 and 6. It is calculable, not only by the technique of single injection of tracer (Chapter 6), but also by the technique of constant infusion. Because volume cleared per unit time is the same for tracer and tracee alike the value obtained with species containing label will be the same as that for unlabeled counterpart solute. Assume that a region or pool containing species a is subjected to constant infusion with labeled species a. For example in Model C of Figure 43 let pool a represent species a in blood which is so infused. But instead of observing the SA the observations will be for concentration of *labeled material in blood* (e.g., units per milliliter) at plateau. Net clearance from blood is calculable regardless of the complexity of the model as a whole, or of the presence of input and output of reversibly connected precursors and products. The derivation of the formula is quite simple. At equilibrium, when plateau concentration of tracer in blood is attained, the rate, r, of input of tracer to blood must equal its rate of loss from blood. Rate of loss is the product of clearance of tracer (Cl_a') and equilibrium concentration at plateau (c_{aE}'). Then,

$$r = Cl_a' \cdot c_{aE}'$$

Substitute clearance of tracee (Cl_a) for equivalent clearance of tracer (Cl_a') and rearrange,

$$Cl_a = r/c_{aE}'$$

Thus,

milliliters of blood cleared/min =

$$\frac{r, \text{ as units of tracer infused/min}}{\text{units of tracer per milliliter blood (at plateau)}}$$

Note that this expression differs from that for disposal rate (Eq. 2) only in that the denominator is expressed as tracer concentration in blood (units of tracer per unit volume) rather than as SA (units of tracer per unit mass).

CHAPTER 10

NONSTEADY STATE

Meaning of Nonsteady State

1. The term *steady state* is sometimes applied to tracer when, for example, it reaches equilibrium either in a closed system or during constant infusion in an open system. But most commonly (as in this discussion) it means a stationary state for unlabeled tracee. If the tracer is an isotope, the condition is sometimes called nonisotopic steady state. The constancy applies to all rates, rate constants, and the amount of unlabeled material in all pools and regions. If any of these change during the period of observation the condition is *nonsteady state*. An obvious example is when the amount of material in the system or a pool increases or decreases because rates of input and output are unequal. These rates could be constant or inconstant while being unequal. Other possibilities come to mind. Input and output, although equal, might simultaneously change in the same direction so as to preserve constancy in quantity of contained material. In such instance the rate constants of output of participating compartments would change. Still another type of nonsteady state is that in which the distribution of output among multiple sites is inconstant. Finally, the quantity, Q, and output rate, F, for a given compartment might change proportionally. In this circumstance the rate constant of output would remain unaltered. For example, assume that both are doubled,

$$k = F/Q = 2F/2Q$$

The usual meaning of a *linear* system will be discussed in Chapter 12.

A Single Pool

GENERAL FORMULAS NOT AVAILABLE FOR COMPLEX SYSTEMS

2. Previous chapters on compartment analysis have included equations applicable to steady-state systems possessing as many as four pools. In a non-steady-state system the formulas for direct calculation become very complex in multicompartment models. If the model were reasonably well character-ized, a solution might be approached via computer simulation. Another alternative is to lump (*judiciously*) several pools into one (Chapter 1). For the glucose pool in the dog Steele (1959) assumed that half of the pool could be considered a single " rapidly reacting " compartment and that this restricted amount of glucose could be used in formulas for single-pool equations. Errors introduced by this kind of simplification are difficult to assess. In the case of glucose another source of error is rapid recycling of the carbon label from nonglucose compounds. In the rat, for example, tracer appears very rapidly in such compounds (Shipley *et al.*, 1967). For the formulas shortly to be derived, it is assumed that only one pool exists, and that it receives no reflux from other pools. A distinction will be made between input and output rates which are either *constant* or *inconstant* while being different from each other.

INTUITIVE PREDICTIONS

3. Nonsteady state is most commonly envisioned as changing pool size caused by unequal rates of input and output. The effect on tracer kinetics is best conceptualized by considering a single pool undergoing constant infusion of tracer (Chapter 9). The pool contains unlabeled solute in continually varying amount, Q, dissolved in water occupying an assumed constant volume. Thus as Q changes the concentration of solute in water (c) changes proportion-ally. Observe changing concentration, c, and specific activity, SA, (tracer units per unit weight of solute) as a function of time. For illustration assume that the system at first is in steady state until after plateau is achieved. The input rate of solute is then abruptly reduced below that of a constant output rate. Intuitively, one may predict that c will fall (thus also Q), and SA will increase. If input rate were increased above output the reverse effects would be expected. Now assume that *input rate* is held *constant* but *output* rate is *reduced*. SA will not change, but c will increase. An increase in output rate likewise will not affect SA, however c will decline. Thus, one would suspect that combined observations for c and SA versus time should identify a rise or fall in either input or output rate.

STATIONARY DISPARATE RATES

Equations

4. The model is that of Figure 48A wherein a single pool receives a constant infusion of tracer. The rates F_1 and F_2 are unequal but constant. Expressions which follow may be converted to those applicable to a single dose (with no constant infusion) by calling infusion rate (r) zero. The differential equation describing rate of change of quantity of tracer (q) in the pool is

$$dq/dt = r - F_2 \cdot \text{SA}(t) \tag{1}$$

Wherein F_2 is constant, but SA changes as a function of time. A companion equation for change in contained quantity of unlabeled tracee (Q) is

$$dQ/dt = F_1 - F_2 \tag{2}$$

This says simply that an incremental or decremental change in Q in unit time is given by the difference between the two constant rates expressed in the same units of time. Turn again to q. At any given point in time indicated by subscripts, t,

$$q_t = Q_t \cdot \text{SA}_t$$

or, as a more general function of time,

$$q(t) = Q(t) \cdot \text{SA}(t) \tag{3}$$

The derivative of Eq. (3) is

$$\frac{dq}{dt} = Q(t) \frac{d(\text{SA})}{dt} + \text{SA}(t) \frac{dQ}{dt} \tag{4}$$

After substituting the right-hand side of Eq. (2) for dQ/dt in foregoing Eq. (4), equate the right-hand side of the latter to the right-hand side of Eq. (1) and rearrange,

$$F_1 = \frac{r - Q(t)[d(\text{SA})/dt]}{\text{SA}(t)} \tag{5}$$

As will be shown later, this equation may be converted to a working formula for calculation of F_1. In its present form it says that the input rate may be

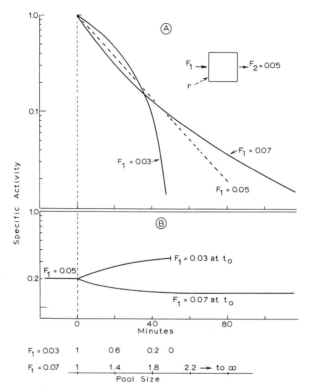

FIG. 48. Curves of SA in a single-pool system (as shown) when ouput rate is held constant at 0.05 units/min, and nonsteady state is brought about by making the input rate greater or less than this (solid lines). In part A a *single* unit dose of tracer is given at zero time. No constant infusion is employed ($r = 0$). In part B *constant infusion* at a rate, r, or 0.01 units/min had yielded a plateau SA of 0.20 prior to zero time. Input rate (F) then was changed, as shown, at zero time.

evaluated by dividing the instantaneous SA into the difference between infusion rate and the product of then existing Q and instantaneous slope of the SA–time curve. For output rate substitute this calculated value of F_1 in Eq. (2),

$$F_2 = F_1 - (dQ/dt) \tag{6a}$$

The last term in Eq. (6a) is the slope of the curve for quantity of unlabeled tracee versus time. Because F_1 and F_2 are considered constant, it is simply the change in Q between any two points in time divided by the associated interval of time. Thus,

$$F_2 = F_1 - (Q_2 - Q_1)/(t_2 - t_1) \tag{6b}$$

If Eq. (1) is restated in terms of instantaneous *quantities* of tracer rather than SA, this permits an integration which yields an expression devoid of the derivative which appears in Eq. (5),

$$dq/dt = r - F_2 q(t)/Q(t) \tag{6c}$$

Consider a reference point in time as zero time when quantity of tracer is q_0 and quantity of solute is Q_0. Then at any time t thereafter,

$$Q(t) = Q_0 + (F_1 - F_2)t \tag{6d}$$

With this substitution in Eq. (6c) integration defines the value of $q(t)$,

$$q(t) = (r/F_1)[Q_0 + (F_1 - F_2)t]$$
$$+ [Q_0]^{+[F_2/(F_1-F_2)]}[Q_0 + (F_1 - F_2)t]^{-[F_2/(F_1-F_2)]}[q_0 - (r/F_1) Q_0] \tag{7a}$$

The equation may be rearranged and stated in terms of specific activities, SA_0 and SA_t, at respective points in time, t_0 and t,

$$\frac{SA_t - (r/F_1)}{SA_0 - (r/F_1)} = \left[\frac{Q_0}{Q_t}\right]^{[F_1/(F_1-F_2)]} \tag{7b}$$

A similar expression may immediately be written in terms of any two points in time, t_1 and t_2, by substituting SA_1 for SA_0, SA_2 for SA_t, Q_1 for Q_0 and Q_2 for Q_t. Also, from Eq. (6b), the quantity $(Q_2 - Q_1)/(t_2 - t_1)$ may be substituted for $F_1 - F_2$ in the exponent on the upper right in Eq. (7b). Together, these substitutions give

$$\frac{SA_2 - (r/F_1)}{SA_1 - (r/F_1)} = \left[\frac{Q_1}{Q_2}\right]^{[F_1(t_2-t_1)/(Q_2-Q_1)]} \tag{8}$$

To apply Eq. (8), time curves of SA and Q would be employed. Values for the latter might be derived from observed concentration of solute in a known constant volume of distribution. The unknown F_1 may be solved for by trial substitution of assumed values until one value is found which makes the two sides of the equation equal (Wall et al., 1957). The value of F_2 follows from the calculated F_1 substituted in Eq. (6b). If r is zero (no constant infusion) and a single dose of tracer is given at or prior to t_1, Eq. (5) may be integrated directly to give

$$F_1 = \frac{(Q_1 - Q_2) \ln(SA_1/SA_2)}{(t_2 - t_1) \ln(Q_1/Q_2)} \tag{9}$$

This also may be put in exponential form as an expression for SA,

$$SA_2 = SA_1 \left(\frac{Q_2}{Q_1}\right)^{[F_1/(F_2-F_1)]} \tag{10}$$

Curves, single dose

5. The model is the single pool of Figure 48A. The illustrative curves of SA are representative of those defined by Eq. (10). The beginning of observation is shown at zero time when quantity of solute (Q_1) is equal to 1, and SA_1 (after a unit dose of tracer) is 1. Beginning at t_0 the output rate remains constant at 0.05 mg/min. For one curve the accompanying input rate is constant at 0.03 mg/min (until the pool is empty at 50 min), and for the other curve the input rate is constant at 0.07 mg/min. Note that as compared to the straight-line simple exponential (dashed line) representing $F_1 = F_2 = 0.05$, the line is convex in one instance and concave in the other. If these were observed curves (Q versus time having been determined via time curves of concentration of solute in a fixed known volume), Eq. (9) would provide a solution for F_1 via any two chosen points in time. Not shown is the effect of holding F_1 at 0.05 and making F_2 either 0.03 or 0.07. The effects on contour would be similar to those when disparity was produced by changing F_1 although the curves would not be identical to those illustrated. If the curve as shown for $F_1 = 0.07$ is observed experimentally an illustrative numerical solution via Eq. (9) is as follows: At times $t_1 = 10$ and $t_2 = 30$, the associated values of Q and SA are $Q_1 = 1.2$; $Q_2 = 1.6$; $SA_1 = 0.53$; $SA_2 = 0.19$, and

$$F_1 = \frac{(1.2 - 1.6)\ln(0.53/0.19)}{(30 - 10)\ln(1.2/1.6)} = \frac{-0.4 \ln 2.79}{20 \ln 0.75}$$

$$= \frac{-(0.4)(1.029)}{(20)(9.71 - 10)} = \frac{-0.41}{-5.8} = 0.071$$

$$F_2 = 0.071 - \frac{1.6 - 1.2}{20} = 0.051$$

The slight error arises from reading values for SA from the curve rather than by using theoretically exact values. Note that this curve resembles one from a steady-state model wherein a concave curve is produced by a reversible exchange between the labeled pool and another pool. Such contour illustrates another uncertainty to add to those attending lumping multiple pools into one as described in Chapter 1. Concavity may be caused either by relatively slow interchange between a system of multiple pools in steady state, or by nonsteady state in a single pool wherein input exceeds output.

Curves, constant infusion of tracer

6. Figure 48B shows constant infusion of tracer at a rate, r, of 0.01 units/min. The pool of size 1 at t_0 had, before this time, been in steady state with $F_1 = F_2 = 0.05$ mg/min. F_1 is abruptly changed at t_0 and held constant thereafter. When reduced to 0.03 mg/min, the SA rises to a potential plateau of r/F_1 or 0.01/

0.03, namely 0.333, but the curve terminates abruptly at this point (50 min) because the pool is empty. When F_1 is increased to 0.07 mg/min the SA approaches r/F_1 at infinity, i.e., $0.01/0.07 = 0.143$. With constant infusion (as contrasted with a single dose) a change in F_2 will not affect specific activity (once plateau is achieved). The plateau preexisting at $0.01/0.05$, i.e., 0.2, is unchanged except that when F_2 is increased to 0.07 and F_1 is held at 0.05 the pool is empty at 50 min. If F_2 is decreased to 0.03 and F_1 held at 0.05 the SA will stay at 0.2 indefinitely. The lack of influence of a change in F_2 is confirmed mathematically by the absence of F_2 in Eqs. (5) and (8). The curves of Figure 48B were constructed via Eq. (7b) with $Q_0 = 1$ at zero time.

RATES DISPARATE AND NOT STATIONARY

Equations

7. If rates F_1 and F_2 are *not constant* in addition to being unequal, i.e., if they are allowed to vary as a function of time, Eq. (2) then embodies $F_1(t)$ and $F_2(t)$, and it is applicable only within a given instant persisting no longer than a brief Δt. Equation (5) is now in terms of $F_1(t)$ in place of F_1. But Eq. (5) remains applicable for a short Δt because F_1 is assumed to be essentially constant within this brief interval. If observed curves for Q and SA are separated into sufficiently short segments, an approximation of F_1 should be possible even though the time span of such segments is longer than the theoretically vanishingly small Δt of calculus (Steele, 1959). The equivalent slope, $d(\text{SA})/dt$, in Eq. (5) is assumed to be that of the straight line between SA_1 and SA_2 at times t_1 and t_2, respectively (which are as close together as data points permit). By the two-point formula this slope is $(\text{SA}_2 - \text{SA}_1)/(t_2 - t_1)$. The value taken for $Q(t)$ is that at the midpoint in time between t_1 and t_2 which approximates the arithmetic mean of Q_1 and Q_2, i.e., $(Q_1 + Q_2)/2$. A similar mean is used for $\text{SA}(t)$ in the denominator. Then Eq. (5) may be approximated in the form

$$F_1(t) \approx \frac{r - [(Q_1 + Q_2)/2][(\text{SA}_2 - \text{SA}_1)/(t_2 - t_1)]}{(\text{SA}_1 + \text{SA}_2)/2} \qquad (11)$$

The values of Q_1 and Q_2 most likely will be obtained indirectly from observed values of concentration. If volume of distribution of solute (V) harboring Q is constant, and is separately measured or known, then $Q = V \cdot c$ (where c is concentration, e.g., milligrams per milliliter). The working equation becomes

$$F_1(t) \approx \frac{r - V[(c_1 + c_2)/2][(\text{SA}_2 - \text{SA}_1)/(t_2 - t_1)]}{(\text{SA}_1 + \text{SA}_2)/2} \qquad (12)$$

Equations (11) and (12) obviously given an *approximation* for F_1 rather than a definitive value. Errors expectedly would increase in proportion to the instability of F_1. Also, as emphasized in Chapter 1, a *pure* single pool having no reversible connections with other pools does not exist in the animal body. Equations applied to a contrived lumped equivalent of a single pool may introduce serious error if interchange between lumped constituents is not sufficiently rapid.

Example, constant infusion

8. Although the curves of Figure 48B are for a system with constant rather than changing values of F_1, Eq. (11) obviously is similarly applicable for numerical approximation. Take the curve with $F_1 = 0.03$ and read to two significant figures when $t_1 = 0$ and $t_2 = 5$ min. Equation (11) gives

$$F_1 = \frac{0.01 - [(1 + 0.9)/2][(0.22 - 0.20)/5]}{(0.2 + 0.22)/2} = 0.0295$$

$$F_2 = F_1 - (dQ/dt) = 0.0295 - (-0.02 \text{ mg/min}) = 0.0495$$

Later in the curve when the slope is flatter the potential error is greater. For example between 40 and 45 min the change in SA is only from 0.321–0.329. A short interval must always be taken if calculated values are to be meaningful when flow rates are unstable. Because random error of the difference between close-set data points is potentially large a smoothed curve provides better reference values.

Example, single dose

9. Equation (11) is adapted to a single dose by making $r = 0$. Take the curve for $F_1 = 0.07$ in Figure 48A between 0 and 5 min

$$F_1 = \frac{-[(1 + 1.1)/2][(0.72 - 1)/5]}{(1 + 0.72)/2} = 0.069$$

Complex Systems

STEWART–HAMILTON EQUATION, CONSTANT DISPOSAL

10. A review of Chapter 5, paragraph 5, will confirm that the derivation of the Stewart–Hamilton equation is not conditional on the assumption that pool size remains constant or that input rate must equal output rate. The fundamental equation for material balance in Figure 28A is

$$D = \int_0^\infty F_2 \cdot \text{SA}(t) \, dt$$

On the left is dose introduced, and on the right is dose lost. What is required is that F_2 *must be constant.* Then

$$D = F_2 \int_0^\infty SA(t)\, dt$$

$$F_2 = D \Big/ \int_0^\infty SA(t)\, dt \tag{13}$$

A similar expression is derivable for rate of irreversible disposal of labeled species a, i.e., DR_a (Chapter 6),

$$DR_a = D \Big/ \int_0^\infty SA_a(t)\, dt \tag{13a}$$

Production rate, PR_a, need not equal DR_a nor must it be constant. The requirement that DR_a must be constant imposes a practical limitation on application of Eq. (13a). The area in the denominator is taken from a curve which must extend for a time sufficiently long that SA is a small fraction of that at onset. It may well happen that DR_a will not be constant for so long. In the special case wherein species a occupies a space behaving effectively as a single rapidly mixed pool inconstancy of DR_a might be suggested by a change in slope, although this also may be caused by altered pool size. A rough appraisal of constancy may be made by smoothing the initial and late segments of the curve into separate simple exponentials. Equation (13a) applied to the first segment (D = full dose) should give a value for DR_a similar to that for the late segment (D = amount of tracer present at onset of segment) (Shipley *et al.*, 1970). A measurement of the change in quantity of unlabeled species a in the system between beginning and end of an observation period will permit calculation of *mean* production rate if it is assumed that equilibration through-out a constant sized space of distribution (e.g., blood and extracellular fluid) is sufficiently rapid that sampling of blood at beginning and end will provide a fair representation of overall change. Let c_{a1} be concentration in blood at beginning of observation to extend for a time span, T, and let c_{a2} represent concentration at the end of this time span. Overall volume of distribution is V. Then, what will be present at the end is what was present at the beginning plus what was added in the interim minus what was lost,

$$V c_{a2} = V c_{a1} + \text{mean } PR_a \cdot T - DR_a \cdot T$$

or

$$\text{mean } PR_a = [V(c_{a2} - c_{a1})/T] + DR_a \tag{13b}$$

An example will illustrate the concept without invoking a formal equation. If V multiplied by respective concentrations shows a decline in content from

10 to 8 mg in 100 min, the mean production rate is 0.02 mg/min less than disposal rate. If disposal rate is 0.1 mg/min the mean production rate is 0.1 − 0.02 = 0.08 mg/min.

SA OF POOLED OUTPUT

11. Described in Chapter 6, paragraph 6, is a method for calculating production–disposal rate of labeled species a by collecting excreted material cumulatively. If collected material is a, itself, the model is unrestricted, but collection of a conversion product requires that the latter shall have no primary precursor other than species a, i.e., all input to the system is via a. Irreversible disposal rate for a, DR_a, is given by the ratio of dose of tracer to the multiple of SA of pooled output and time of collection, T. The latter must be sufficiently long to recover essentially all tracer destined to leave at the chosen collection "site." This method is applicable in a certain type of nonsteady-state system. It will give weighted mean rate of DR_a when this rate is inconstant provided fluctuations in overall DR_a are matched by proportional changes in output rate of the species collected. For example, in Figure 29, if output from b is being collected the ratio F_{ob}/DR_a must be constant at all times. Proof of the formulation will begin with a steady-state system. In such a state, as pointed out in Chapter 6, the foregoing ratio is the fraction of unlabeled tracee of species a which is excreted as b, and it is matched by fraction of dose of tracer placed in a which is lost via species b. Thus, in steady state,

$$F_{ob}/DR_a = (\text{tracer excreted as } b)/D$$

If the two rates on the left fluctuate in constant ratio the expression may be written in terms of weighted means for each,

$$\text{mean } F_{ob}/\text{mean } DR_a = (\text{tracer excreted as } b)/D \qquad (14)$$

The ratio on the right also is unaffected by proportional changes in rates. Rearrangement of Eq. (14) gives

$$\text{mean } DR_a = D(\text{mean } F_{ob})/\text{tracer excreted as } b$$

Multiply numerator and denominator each by T. Above the line the resulting product, mean $F_{ob} \cdot T$, represents mass of unlabeled tracee excreted, while the other T is retained in the denominator,

$$\text{mean } DR_a = D(\text{tracee excreted as } b)/(\text{tracer excreted as } b)T \qquad (15)$$

The ratio of the values in parenthesis is the inverse of α_b, the SA of accumulated material. Thus,

$$\text{mean } DR_a = D/\alpha_b T \qquad (16)$$

As in the preceding paragraph, PR_a also is calculable by difference if any change in the existing amount of species a in the system can be measured. In one circumstance this type of equation is applicable when members of a group of excreted products do not bear constant ratios to total disposal rate. This is when *all* excreted material is collected and pooled for joint SA. In other words, for Figure 29 collect a, b, and c *together* and measure joint SA. Deserving comment is the conclusion that the equation for cumulative output is applicable in the face of an inconstant disposal rate whereas the Stewart–Hamilton equation is not. This is related to the fact that the latter is based solely on a curve which reflects ultimate *loss* of tracer, whereas the former is based on a value for SA arising via *accumulation* of tracer diluted by attending unlimited accumulation of tracee.

CONSTANT CLEARANCE, INCONSTANT DISPOSAL RATE

12. Assume that DR_a, the irreversible disposal rate of species a from a pool or region which it occupies (e.g., blood) is not constant. Production rate (PR_a) likewise may vary with time. Thus, the concentration of a in blood (c_a) and consequently the amount in blood (Q_a) can vary with time in any unspecified manner including cyclic fluctuation. Sites of input and output to the overall system with which blood is connected are not restricted (Figure 32B). The proportional loss via various products need not be constant. The important qualifying assumption is that overall *clearance* of species a from blood is *constant*. The aim is to calculate the mean DR_a and PR_a during a given period of observation. DR_a, in terms of clearance (Cl_a) is $Cl_a \cdot c_a$ (see Chapter 6, Eq. (21b)). The *amount* of species a lost by irreversible disposal during a short segment of time, Δt, is $Cl_a \cdot c_a \cdot \Delta t$, and the amount lost between two more widely separated points in time, t_1 and t_2, is the summation of bits represented by a series of such triple products. Thus,

$$\text{amount of species } a \text{ lost in the interval from } t_1 \text{ to } t_2 = \int_{t_1}^{t_2} Cl_a \cdot c_a(t)\, dt$$

With clearance assumed constant Cl_a may be placed in front of the integral. Division by the time span, $t_2 - t_1$, then gives mean DR_a as quantity disposed of per unit time,

$$\text{mean } DR_a = Cl_a \int_{t_1}^{t_2} c_a(t)\, dt \Big/ (t_2 - t_1) \tag{17}$$

The curve for concentration of species a in blood, $c_a(t)$, may be effectively integrated by graphic estimate of area subtended between t_1 and t_2 or by the trapezoid rule. Clearance may be measured and shown to be constant in the face of varying blood concentrations by tracer methods given in

Chapter 6, paragraph 17, or Chapter 9, paragraph 16. If, at the end of the period of observation, c_a is essentially the same as at the beginning and volume of distribution is assumed constant, the value for production rate (PR_a) can be considered the same as the calculated disposal rate. (See Kowarski *et al.*, 1971). If the end value for c_a differs from the starting value and volume of distribution, V_a, is known, PR_a is calculable as a different rate as explained in paragraph 10.

13. Recognizable in Eq. (17) is a moiety which represents concentration of solute between t_1 and t_2. Thus,

$$\text{mean } c_a = \int_{t_1}^{t_2} c_a(t) \, dt \bigg/ (t_2 - t_1)$$

The fraction on the right is simply area under the curve divided by the time span. This permits Eq. (17) to be restated in another form,

$$DR_{a(\text{mean})} = Cl_a \cdot c_{a(\text{mean})} \tag{17a}$$

or

$$c_{a(\text{mean})} = DR_{a(\text{mean})}/Cl_a \tag{18}$$

Equation (18) would be useful to assess mean concentration of species a when DR_a has been measured by another method (e.g., SA of cumulated output).

14. A special adaptation of Eq. (13b) permits estimation of mean PR_a for short intervals of time during which a change in concentration of species a in blood can be approximated by a straight line (in linear coordinates) between c_{a1} and c_{a2}. If clearance is constant, the value of DR_a in the equation becomes

$$\text{mean } DR_a \approx Cl_a \left(\frac{c_{a1} + c_{a2}}{2} \right)$$

Also, if species a, e.g., that in blood, is so rapidly equilibrated with secondary compartments that the system may effectively be treated as a single pool, and there exists an effective unchanging single rate constant of disposal (k) for the system in composite, then

$$\text{mean } DR_a = k Q_{a(\text{mean})} \qquad \text{or} \qquad \text{mean } DR_a \approx k \left(\frac{c_{a1} + c_{a2}}{2} \right) V$$

Equation (13b) for mean PR_a during short time, T, becomes

$$\text{mean } PR_a \approx V \left[\frac{c_{a2} - c_{a1}}{T} + \frac{k(c_{a1} + c_{a2})}{2} \right] \tag{19}$$

Weitzman *et al.* (1971) have used Eq. (19) to calculate short term bursts of cortisol production. However, the decay curve for labeled cortisol is not a simple exponential function wherein k appears as a single slope of an effective single pool. At least two slopes exist, and the terminal one, from which these investigators determine nominal half time (to give k), is delayed for $\frac{1}{2}$–1 hr.

CHAPTER 11

CIRCULATION RATE MEASURED BY TISSUE SATURATION OR DESATURATION

A Single Homogeneous Tissue

THE MODEL

1. The simple block in Figure 49A represents homogeneous tissue supplied by a capillary bed shown diagrammatically as a single tube. The block of tissue, with its associated vascular input and output, is not unlike a single-compartment system undergoing exponential washout as described in Chapter 1. The essential difference is that fluid which enters via arterial blood does not mix *instantaneously* with tissue fluid. All mixing, either of water or tracer, is by diffusion through membranes. Nevertheless, as will be shown later, if the tissue is enriched with a diffusible tracer, and the time course of loss is recorded, a rate constant for decay can be formulated which is directly related to the rate of flow of perfusing blood. Determination of perfusion rate by such an approach is known as a *desaturation* method. Before considering such a method in detail a *saturation* method will be examined. In a saturation experiment the tracer is delivered continually over an extended time via arterial blood. The tissue slowly imbibes the tracer via diffusion. This loss (preceding ultimate equilibration) results in a lower concentration in venous effluent than in arterial input. The A–V difference, as plotted over an extended time, is the basis for the calculation. Kety, who developed the method to determine cerebral blood flow, has reviewed the theory and basic formulations (Kety, 1951 and 1960). The tracer originally employed was nitrous oxide delivered to arterial blood by prolonged inspiration of the gas. Because of greater ease of measurement in blood radioactive gases such as [85]Kr or ^{133}Xe have largely supplanted N_2O.

CALCULATIONS BY A–V DIFFERENCE

Formulas for saturation phase

2. The amount of tracer existing at any point in time in the mass of tissue in Figure 49A will be the difference between what has been delivered by arterial blood up to that time, and what has been removed by venous blood during

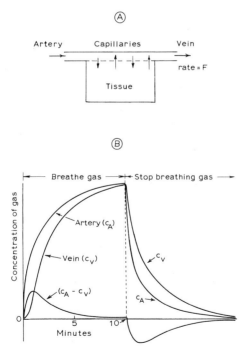

FIG. 49. A model for perfusion of a single homogeneous tissue (Part A), and curves for arterial concentration, venous concentration, and A–V difference during saturation and desaturation (Part B).

the same period of time. Let the symbol q represent the amount of tracer present in tissue at any given time. An expression will be derived to predict the value of q from the time curves of respective concentrations of tracer in arterial and venous blood. Such curves are shown in Figure 49B. From zero time when inspiration of gas is begun, the arterial concentration will rise from zero to an eventual plateau value at which time all body tissues, including the sample under observation, will be in equilibrium with tracer in both arterial and venous blood. Before that time, while tracer is moving to tissue, the venous concentration will be lower than arterial concentration. Although blood concentrations are not constant with time they may be considered constant during a very short time span, Δt. First, imagine that arterial blood having tracer concentration, c_A, delivers all its tracer to tissue. The amount will be rate of blood flow, F, multiplied by arterial concentration of tracer, c_A, multiplied by the time span, Δt. But what is delivered to tissue is partially offset by what is simultaneously lost to venous blood. The amount lost during this same span of time is the triple product of F, venous concentration, c_V, and Δt. This must be subtracted to give the net accumulated quantity of tracer,

$$\text{accumulation during } \Delta t = F \cdot c_A \cdot \Delta t - F \cdot c_V \cdot \Delta t$$

or

$$\text{accumulation during } \Delta t = F(c_A - c_V)\, \Delta t$$

Now consider a prolonged period between zero time and any subsequent point in time to be called ω (omega). The amount present at t_ω will be designated as q_ω. It will be the sum of the series of bits yielded during preceding serial segments of time of length Δt,

$$q_\omega = \sum_\omega F(c_A - c_V)\, \Delta t$$

This is represented by the integral from t_0 to t_ω,

$$q_\omega = \int_{t_0}^{t_\omega} F(c_A - c_V)\, dt \tag{1a}$$

Flow rate, F, being constant, is placed outside the integral,

$$q_\omega = F \int_{t_0}^{t_\omega} (c_A - c_V)\, dt \tag{1b}$$

or

$$F = \frac{q_\omega}{\int_{t_0}^{t_\omega} (c_A - c_V)\, dt} \tag{1c}$$

Equation (1a) also follows directly from the differential equation for rate of change of q,

$$dq/dt = F[c_A(t) - c_V(t)] \tag{2}$$

3. The lowermost curve of Figure 49B is the derived curve for A–V difference. The area under this curve between t_0 and t_ω is the denominator of Eq. (1c). In the present instance t_ω is 10 min. Rate of blood flow through the mass of tissue would be calculable if the amount of tracer in tissue (q) were known at 10 min. This not ordinarily being known, an equivalent expression for q_ω must be substituted in the numerator of Eq. (1c). In making this modification of the equation an important assumption is that the low solubility of the tracer gas in water and in tissue substance, and its rapid rate of diffusion, foster a rapid equilibration between blood and tissue. Consequently, as blood leaves capillaries for venous channels the equilibration between the two sites may be considered complete. Thus, assume tentatively that tissue concentration (c) equals venous concentration (c_V). However, the solubility of the gas in tissue is likely to be different from that in blood. The partition coefficient (λ) must be introduced to correct for this differential at equilibrium. For example, if an *in vitro* trial with the tissue in question shows a tissue/blood ratio of 1.2 units of gas per gram of tissue to 1 unit per milliliter of blood, the partition coefficient is 1.2. Therefore, at equilibrium *in vivo* $c = 1.2\, c_V$. More generally,

$$c = \lambda c_V \tag{3a}$$

or,

$$c_V = c/\lambda \tag{3b}$$

Note that the partition coefficient is in terms of *weight* units for tissue versus *volume* units for equilibrating blood. Conventionally these are grams and milliliters. Thus when c_V in Eq. (3a) is expressed as tracer units per milliliter the multiplication by λ gives tracer units per *gram* of tissue, which by definition is c. If mass of tissue is called m, $c = q/m$ and Eq. (3b) may be written

$$c_V = \frac{q/m}{\lambda} \tag{4}$$

or

$$q = m\lambda c_V \tag{5}$$

Let the venous concentration, c_V, in this expression, when taken at time t_ω, be called $c_{V\omega}$. The triple product on the right, then, is equal to q_ω. It may be substituted for the latter in Eq. (1c),

$$F = \frac{m\lambda c_{V\omega}}{\int_{t_0}^{t_\omega} (c_A - c_V)\, dt} \tag{6}$$

This is not yet a working equation unless tissue mass, m, is known. However the rate of flow per *unit* mass is calculable without knowing the total weight of tissue. Rate per unit mass is F/m. Call this f. Then, dividing the right side likewise by m,

$$f = \frac{\lambda c_{V\omega}}{\int_{t_0}^{t_\omega} (c_A - c_V)\, dt} \tag{7}$$

4.   Equation (7) is the conventional working formula. If units of tracer concentration are per milliliter of blood, and time is in minutes, the rate, f, is milliliters per minute per gram of tissue. In Figure 49B the area under the lowermost curve to 10 min, which provides the denominator of Eq. (7), might be measured by planimetry, or by the trapezoid rule (Chapter 5, paragraph 6). The 10 min value for c_V would represent $c_{V\omega}$ in the numerator. For cerebral blood flow the concentration, c_A, is measured from any artery, and c_V from a catheter in the internal jugular vein. Theoretically the time span to t_ω in Eq. (7) can be of any unspecified duration if the tissue is homogeneous as to λ and f in all of its parts. Brain does not qualify in this respect because the partition coefficient and perfusion rate are not the same for white and gray matter. For inhomogeneous tissue Eq. (7) provides a fair estimate of *mean* rate of flow for the entire complex only if t_ω is sufficiently long for c_V to approach c_A (Sapirstein and Ogden, 1956; Lassen and Klee, 1965). Ten minutes usually suffices for a normal brain, but not in the presence of lesions which cause local reduction in perfusion rate. This point will be raised again later.

Formulas for the desaturation phase

5.   Tracer in the form of poorly soluble *gas* is advantageous for observing washout from tissue because arterial blood, which ideally should be free of tracer while it is cleansing the presaturated tissue, is rendered nearly tracer-free during a single pulmonary passage. Note in Figure 49B that arterial concentration falls very abruptly when delivery of gas is stopped. The arterial and venous curves during washout may be used to calculate f via a modified form of Eq. (7). Instead of assuming that the first point of observation is for zero concentration in blood and tissue, the equation is made more general. Equation (1b) may easily be altered to give quantity of tracer *present* at the end of observation instead of quantity *accumulated*. Thus,

$$q_2 = q_1 + F \int_{t_1}^{t_2} (c_A - c_V)\, dt \tag{8}$$

The quantity q_1 is present at t_1 and q_2 is present at t_2. This says simply that what is present is that which existed at t_1 plus that which accumulated between

t_1 and t_2. Note in Figure 49B that during washout c_V exceeds c_A. Therefore the value in the integral becomes negative (loss of tracer) and the amount existing at t_2 is q_1 minus what is lost. When an amount equal to q_1 is lost $q_2 = 0$. Substitution in a manner similar to that for Eqs. (6) and (7) gives

$$f = \frac{\lambda(c_{V2} - c_{V1})}{\int_{t_1}^{t_2} (c_A - c_V) \, dt} \tag{9}$$

In parentheses in the numerator are venous concentration at t_1 and t_2. In applying this formula to desaturation curves, c_{V1} would ordinarily be the point marking onset of the desaturation phase and c_{V2} would be a point later in time, ordinarily when the two concentrations both approach a similarly low value. Equation (9) theoretically applies to any segment of a saturation or desaturation curve, or a segment which includes both phases. Note that when c_{V1} and t_1 are zero, Eq. (9) becomes Eq. (7). Also note that if Eq. (9) is applied to the interval marked by beginning saturation to complete desaturation (theoretically at $t_2 = \infty$), it will give $f = 0$. The equation is useless. Its sensitivity is diminished if saturation and desaturation phases are used jointly. Equation (9) is most profitably applied to the pure desaturation phase. Restrictions applying to a nonhomogeneous organ will be discussed later under this heading.

Equations for tissue concentration

6. For certain purposes it may be useful to have equations for concentration of tracer in tissue (c) as a function of time. Equation (2) may be altered by substituting c/λ for c_V (see Eq. 3b). Then Eq. (2) becomes

$$\frac{dq}{dt} = F\left(c_A - \frac{c}{\lambda}\right) \qquad \text{or} \qquad \frac{dq}{dt} = \frac{F}{\lambda}(\lambda c_A - c)$$

Conversion to an expression for tissue concentration is accomplished by dividing by m (i.e., tissue mass),

$$\frac{dc}{dt} = \frac{F}{\lambda m}(\lambda c_A - c) \tag{10}$$

Note that the fraction preceding the parentheses is constant. Let this term be called k, where

$$k = F/\lambda m \tag{11}$$

Then by substitution in Eq. (10),

$$dc/dt = k(\lambda c_A - c) \tag{12}$$

This may be integrated for three separate boundary conditions. First, when c_A is instantaneously raised to a constant level at zero time, the time curve for c of tissue thereafter will be

$$c = \lambda c_A (1 - e^{-kt}) \tag{13}$$

If c_A, once established, suddenly is allowed to go to zero so as to observe the washout phase, Eq. (12) becomes,

$$dc/dt = -kc$$

Integration yields a simple exponential decay curve,

$$c = c_0 e^{-kt} \tag{14}$$

c_0 is tissue concentration at the beginning of observation during washout. If c_A is changing as a function of time in a nonspecified manner except that it is zero at zero time,

$$c = \lambda k e^{-kt} \int_{t_0}^{t_\omega} c_A e^{+kt} \, dt \tag{15}$$

If k is known via the parameters of Eq. (11), the value for the quantity in the integral of Eq. (15) can be determined graphically as area under a curve of the product of the arterial concentration and the exponential term at serial points in time between t_0 and t_ω.

A Nonhomogeneous Organ

By A–V Difference

The model

7.   In the model of Figure 50 the organ consists of tissue a with mass m_a and partition coefficient λ_a, along with tissue b having a different mass m_b and a different partition coefficient λ_b. The separate channels of venous output have tracer concentrations denoted as c_{Va} and c_{Vb}, respectively. These combine in a common collecting vein to give a concentration of c_V after mixing. Arterial concentration is c_A, as before; quantities of tracer in the respective tissues at a given time are q_a and q_b at concentrations c_a and c_b (which also are q_a/m_a and q_b/m_b). Rates of flow, F_a and F_b, are nonidentical, and when combined will give an overall rate through the pair of tissues of F ml/min. The goal is to determine f_{mean}, the *mean* rate of flow for the combined tissues as milliliters per minute per gram.

Saturation phase

8. Equation (1c) may be modified to embrace both tissues by recognizing two sites which harbor tracer in respective amounts q_a and q_b at the end of observation (symbol ω),

$$F = \frac{q_{a\omega} + q_{b\omega}}{\int_{t_0}^{t_\omega} (c_A - c_V)\, dt}$$

In the numerator substitute for tracer content in each tissue equivalent expressions stated, as per Eq. (5), in terms of separate concentrations in respective venous effluent channels,

$$F = \frac{m_a \lambda_a c_{Va} + m_b \lambda_b c_{Vb}}{\int_{t_0}^{t_\omega} (c_A - c_V)\, dt} \tag{16}$$

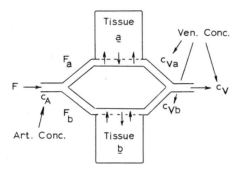

FIG. 50. A model for perfusion of two tissues having different perfusion rates, $F_a = 2$ ml/min and $F_b = 8$ ml/min.

Because of its algebraic configuration this equation cannot be converted to a simple expression for mean rate of flow per gram (f_{mean}) as was the case for Eq. (7) where a *single m* was encountered and F/m (i.e., f) was defined on the left. Mean flow rate in the present system is $F/(m_a + m_b)$, but the two masses cannot be taken from the numerator of Eq. (16) as a sum to provide such a denominator for F on the left. If each mass were independently known and the *separate* venous concentrations were monitored, Eq. (16) would be usable. But a simplifying modification is possible if termination of observation is at infinite time (t_∞) instead of at an unspecified point, t. In the present instance t_∞ will mean a duration *sufficiently long* that tracer concentration in *all* blood and *both* tissues is essentially the same. Because of this identity the value for mixed venous blood at that time ($c_{V\infty}$) may be substituted for both c_{Va} and

c_{Vb} in the numerator of Eq. (16), which then becomes $c_{V\infty}(m_a \lambda_a + m_b \lambda_b)$. Then we may write

$$\frac{F}{m_a \lambda_a + m_b \lambda_b} = \frac{c_{V\infty}}{\int_{t_0}^{t_\infty} (c_A - c_V)\, dt} \tag{17}$$

Equation (17) may be simplified for practical application if the two partition coefficients are *equal* and are simply represented by λ, i.e., $\lambda_a = \lambda_b = \lambda$. The denominator on the left becomes $\lambda(m_a + m_b)$ and for mean rate of flow through the composite mass of two tissues having equal partition coefficients, we may write

$$f_{\text{mean}} = \frac{F}{m_a + m_b} = \frac{\lambda c_{V\infty}}{\int_{t_0}^{t_\infty} (c_A - c_V)\, dt} \tag{18}$$

The duration of time to follow c_A and c_V so that c_{Va} and c_{Vb} are predicted to be equal is identified by that time when arterial and pooled venous concentrations approach equality. This point in time marks the approach of general equilibrium. The necessity for extending observations beyond any arbitrary t_ω can be appreciated by realizing that the concentration change in tissue with more rapid perfusion dominates Eq. (18) during early time whereas change in tissue having slow perfusion is dominant at later time. Because venous concentration reflects tissue concentration, the contribution of rapidly perfused tissue to the value of c_V of Eq. (18) would be dominant if observations were only for a brief period. Sapirstein and Ogden (1956) have described the behavior in terms of exponential functions.

9. Equation (18) is also applicable when the two partition coefficients are unequal, provided λ represents a *weighted mean*, i.e., when it has been weighted for the proportional masses of the substituent tissues. Thus for the weighted mean λ,

$$\lambda_{\text{mean}} = \lambda_a \left(\frac{m_a}{m_a + m_b}\right) + \lambda_b \left(\frac{m_b}{m_a + m_b}\right)$$

The weighted partition coefficient is calculable if the separate partition coefficients and mass ratios are known. It also may be determined simply by measuring the coefficient experimentally for the *entire* organ. For brain the value of λ for the organ as a whole is approximately 1.08 with both xenon and krypton.

Desaturation phase

10. As was the case for the saturation phase a simple working equation evolves if observations are continued for a time sufficiently long to be of nominal infinite duration. The time span will be from full saturation at equilibrium (t_1) until concentration in common venous effluent is near zero (t_∞). An expression for two tissues counterpart to Eq. (8), which was for one tissue, has symbols $(q_a + q_b)_1$, and $(q_a + q_b)_\infty$ corresponding to amount of tracer present at the foregoing respective times,

$$(q_a + q_b)_\infty = (q_a + q_b)_1 + F \int_{t_1}^{t_\infty} (c_A - c_V)\, dt \tag{19a}$$

or

$$F = [(q_a + q_b)_\infty - (q_a + q_b)_1] \Big/ \int_{t_1}^{t_\infty} (c_A - c_V)\, dt \tag{19b}$$

As in Eq. (16), the amount of tracer may be expressed as the triple product of mass, partition coefficients, and concentration in local effluent,

$$F = [(m_a \lambda_a c_{Va} + m_b \lambda_b c_{Vb})_\infty - (m_a \lambda_a c_{Va} + m_b \lambda_b c_{Vb})_1] \Big/ \int_{t_1}^{t_\infty} (c_A - c_V)\, dt \tag{19c}$$

At t_1 (full saturation) all concentrations are equal. Therefore the venous concentration of common effluent (c_V) may be substituted for c_{Va} and c_{Vb} in the parentheses on the right side of the numerator. And because c_{Va} and c_{Vb} approach zero at infinity all values in the parentheses on the left side become zero. Then at t_1, with the common venous concentration on the right called c_{V1},

$$F = -c_{V1}(m_a \lambda_a + m_b \lambda_b)_1 \Big/ \int_{t_1}^{t_\infty} (c_A - c_V)\, dt \tag{19d}$$

If a *weighted* mean partition coefficient, λ_{mean}, is used, Eq. (19d) may be simplified to give a working expression resembling Eq. (18) except that it applies to tracer loss (desaturation) rather than saturation,

$$f_{mean} = -\lambda_{mean} c_{V1} \Big/ \int_{t_1}^{t_\infty} (c_A - c_V)\, dt \tag{20}$$

Extrapolation

11. To avoid undue prolongation of the breathing period in a saturation study an extrapolation can be performed to estimate both the numerator and denominator at infinity in Eq. (18). The near plateau in arterial concentration which, in brain studies, ordinarily is approached in 10–15 min, approximates the venous concentration ultimately to be attained at equilibrium. Hence the value of c_A at near-plateau may be substituted for $c_{V\infty}$. The curve for A–V difference in the denominator has a tail which approximates a simple exponential decay curve approaching zero at infinity. Consequently, its terminal portion may be extrapolated to infinity on a semilog plot for estimation of area to infinity (Chapter 5). (See also Lassen and Klee, 1965.) A pitfall to remember is the effect of a substantial mass of abnormal tissue with very poor perfusion rate. This can postpone the development of the *true* terminal slope and render its recognition difficult, particularly if multiple local rates constitute a broad spectrum. This applies also to the desaturation phase (Eq. 20). Note also that Eq. (20) is accurate only if t_1 is at a point marking a sufficiently long period to achieve essentially complete saturation of all tissues. For two tissues at t_1 the following equality is assumed: $c_{Va} = c_{Vb} = c_V = c_A$.

External Monitoring

A SINGLE HOMOGENEOUS TISSUE

12. External monitoring with a radiation detector during washout of a radioactive gas (desaturation) may be employed to evaluate rate of blood flow through tissue. The concept is best approached in terms of the kinetics of removal of tracer from a pool (Chapter 1). Equation (14) is the predicted expression for tissue concentration versus time. This equation may be rearranged to give an expression for *fractional* change,

$$c(t)/c_0 = e^{-kt} \qquad (21a)$$

The units now are in terms of fraction of concentration existing at zero time (beginning desaturation). Multiplying the numerator and denominator on the left by tissue mass converts this expression to one for *quantity* of tracer,

$$q(t)/q_0 = e^{-kt} \qquad (21b)$$

or, if a radiation detector is viewing the tissue, a common conversion factor above and below on the left converts the units to detector readings, R,

$$R(t)/R_0 = e^{-kt} \qquad (21c)$$

R_0 is the detector reading at zero time (beginning desaturation after tissue loading), and $R(t)$ represents the instantaneous detector reading at any time, t, as time progresses thereafter. The important element common to all the foregoing expressions is the exponential slope, k. It has several important equivalencies. The derivation of Eq. (11) defined it as

$$k = F/\lambda m \qquad (22a)$$

where F is rate of blood flow through the entire tissue mass, m. Note that the moiety F/m is rate of flow per unit mass, i.e., f. Thus, by arranging Eq. (22a) so that F/m appears as the numerator and is replaced by f,

$$k = f/\lambda \qquad (22b)$$

Still another form is possible. As explained in paragraph 3, the partition coefficient, λ, is the ratio of tracer concentration in tissue to that in blood at equilibrium. Venous blood is considered to be in equilibrium with tissue. Query: What volume of water in tissue would be required to harbor tracer in the same concentration as that of venous blood? This will be called V. It may be considered a "volume of distribution." This volume will be given by the classical isotope dilution formula $V = q/c_V$. Equation (5) defines q as being equal to $m\lambda c_V$. Substituting the latter for q in this dilution formula,

$$V = \lambda m \qquad (22c)$$

Substitute V for λm in Eq. (22a) and another expression for k emerges,

$$k = F/V \qquad (22d)$$

This is the form of the classical rate constant for tracer washout from a single liquid pool (Chapter 1).

13.   When exponential slope is defined as in Eq. (22b), Eq. (21c) may be written with f/λ in place of k,

$$R(t)/R_0 = e^{-(f/\lambda)t} \qquad (23)$$

If the partition coefficient is known, this observed washout curve gives rate of flow per unit weight of tissue in very simple fashion by graphical estimate of the exponential slope. Then Eq. (22b) gives

$$f = k\lambda \qquad (24)$$

A single homogeneous tissue may first be saturated by breathing radioactive gas, or presaturation may be accomplished by injecting the dissolved gas into the afferent artery as a bolus, or (as in muscle) the dissolved gas may be injected directly into tissue. A radiation detector over the tissue yields a decay curve for washout in which serial values are expressed as fraction of peak value (Eq. 21c). If flow rate is not constant during the period of washout the slope of the curve will vary accordingly. Nevertheless, the instantaneous slope at any point will reflect the existing rate at that time. With a gas such as ^{133}Xe the desired equilibrium between tissue and venous effluent is attained rapidly enough (as contrasted with an ion such as ^{24}Na) that even at high rates of blood flow a change in rate is rather promptly reflected by a change in slope (Lassen, 1964). Instantaneous slope is estimated by drawing a tangent to the curve on a semilog plot. The slope of the tangent is estimated graphically (Chapter 1). In that gas is almost totally cleared from blood during pulmonary passage, a correction for recirculation is unnecessary when a small amount is injected locally. The amount recirculating following a dose given intra-arterially may be sufficient to justify a slight correction. The magnitude of the correction can be determined by injecting the same dose of gas intravenously and monitoring for the time curve over the experimental site being observed. Express points on this curve as fraction of the peak after arterial injection and subtract from corresponding time points on the experimental curve following arterial injection.

TWO OR MORE DISSIMILAR TISSUES

Calculation of separate flow rates and overall mean

14. Consider that separate tissues a and b of Figure 50 are each first saturated with separate amounts of tracer. That in tissue a at zero time is q_{a0}, and that in tissue b is q_{b0}. Equation (14), modified for quantity (rather than concentration) gives the following for each tissue,

$$q_a = q_{a0}e^{-k_at}, \qquad q_b = q_{b0}e^{-k_bt}$$

An external counter, viewing the tissues conjointly, would respond to the sum of these two functions which when added together are expressed by the following equation for a complex exponential function,

$$q_a + q_b = q_{a0}e^{-k_at} + q_{b0}e^{-k_bt} \tag{25}$$

The observed curve may be peeled solely to determine the exponential slopes k_a and k_b. Then, if the separate partition coefficients are known and can be *matched with appropriate* slopes, Eq. (24) will give the separate rates of flow per gram,

$$f_a = \lambda_a k_a, \qquad f_b = \lambda_b k_b \tag{26}$$

If, in addition, the ratio of masses of separate tissues is known, a value emerges for weighted mean rate of flow per gram of composite mass,

$$f_{mean} = f_a\left(\frac{m_a}{m_a + m_b}\right) + f_b\left(\frac{m_b}{m_a + m_b}\right) \tag{27}$$

Knowledge of *actual* numerical weights of separate tissues gives flow rates in terms of volume per unit time to each separate mass in its entirety,

$$F_a = m_a f_a, \qquad F_b = m_b f_b \tag{28}$$

Direct calculation of mean flow per unit mass

15. Without knowing the ratio of masses the weighted mean rate per unit mass of two combined dissimilar tissues is calculable from a curve obtained by external monitoring if tracer gas is delivered to the tissues as a slug in an arterial bolus, and assumptions are made concerning tissue content at onset of desaturation (Høedt-Rasmussen *et al.*, 1966). Figure 50 will illustrate the principle. A small bolus is abruptly delivered to the main artery. The fraction of dose moving to *a* in a given interval of time will be proportional to the respective rates of flow: $\frac{2}{10}$ to *a*, and $\frac{8}{10}$ to *b*. Assume that the partition coefficients may be neglected because the two fractions of the short-lived bolus deliver essentially all of their contained tracer to the respective tissues. The tissues act as sinks for all the tracer delivered during such a short interval. Thus, the fraction of dose to *a* is 0.2, and fraction of dose to *b* is 0.8. The plotted curve obtained by external monitoring will have an initial sharp peak followed by a slower decline which represents washout of tracer. Let the peak be taken as zero time at which time the detector reading will be R_0. Readings thereafter along the curve are $R(t)$. If the latter are expressed as ratio to peak, the declining curve will have the following equation,

$$R(t)/R_0 = 0.2e^{-k_a t} + 0.8e^{-k_b t} \qquad (29a)$$

The ratio on the left is the same as the ratio of quantity of tracer in the combination of tissues at any given time to the amount present in the beginning. The coefficients, 0.2 and 0.8 on the right are the respective fractions of the dose allocated to the separate tissues. This is a modification of Eq. (25), which did not specify that the coefficients represented two component fractions of a single abruptly delivered dose. At zero time the value of the left-hand side of Eq. (29a) is equal to 1, which is the peak value (R_0/R_0). If the curve is peeled, the respective coefficients will appear as two intercepts, I_a and I_b. The general equation will be

$$R(t)/R_0 I = {}_a e^{-k_a t} + I_b e^{-k_b t} \qquad (29b)$$

Because the intercepts will be in the same ratio to each other as the ratio of dose delivered to the tissues, which in turn is the same as the ratio of F_a to F_b, then

$$\frac{I_a}{I_b} = \frac{F_a}{F_b} = \frac{f_a m_a}{f_b m_b} \qquad (30)$$

From this relationship,

$$\frac{m_a}{m_b} = \frac{I_a f_b}{I_b f_a} \qquad (31)$$

Equation (27) may be modified by dividing the numerator and denominator in parentheses on the left by m_a and on the right by m_b,

$$f_{\text{mean}} = f_a \left[\frac{1}{1 + (m_b/m_a)} \right] + f_b \left[\frac{1}{1 + (m_a/m_b)} \right]$$

Substitute the right-hand side of Eq. (31) for the mass ratios above, and simplify to get the following,

$$f_{\text{mean}} = \frac{1}{(I_a/f_a) + (I_b/f_b)} \tag{32}$$

As in the preceding section this is an effective working equation only if, in applying Eq. (24) to derive the separate values for f_a and f_b, the observed peeled slopes can be matched with the appropriate distribution coefficients.

MODIFIED STEWART–HAMILTON EQUATION

Single homogeneous tissue

16. Expression n of Chapter 7, paragraph 11, when applied to a single pool having units of volume , V, gives rate of volume flow (e.g., milliliters per minute) through the pool. Thus,

$$F = \frac{V}{\int_0^\infty [R(t)/R_0]\, dt} \tag{33a}$$

In the case of a single homogeneous tissue, the size of V is that effective nominal volume within the tissue which may be designated to contain tracer in the same concentration as that of venous effluent. Thus, via Eq. (22c), $V = \lambda m$. Therefore,

$$F = \frac{\lambda m}{\int_0^\infty [R(t)/R_0]\, dt} \tag{33b}$$

or, transposing the m to give F/m on the left,

$$f = \frac{\lambda}{\int_0^\infty [R(t)/R_0]\, dt} \tag{33c}$$

$R(t)/R_0$ is the time curve for the ratio of serial detector readings to the initial reading at beginning of desaturation. In a single homogeneous tissue, saturation may have been accomplished by inhaling the gas, or by local or arterial injection. Note that Eq. (33c) is actually the same as Eq. (24). $R(t)/R_0$ is the curve, e^{-kt}, the indicated integral of which has the value $1/k$.

Two or more dissimilar tissues

17. All that is required to convert Eq. (33a) to an expression for multiple dissimilar tissues is to modify the numerator. For the two tissues in Figure 50, for example, the volume which is designated to contain tracer in the same concentration as joint venous effluent is total mass multiplied by the *weighted mean* distribution coefficient. Thus,

$$F = \frac{(m_a + m_b)\lambda_{mean}}{\int_0^\infty [R(t)/R_0]\, dt} \tag{34a}$$

Transposing $m_a + m_b$ to the left gives $F/(m_a + m_b)$, which is weighted mean flow rate per gram (f_{mean}),

$$f_{mean} = \frac{\lambda_{mean}}{\int_0^\infty [R(t)/R_0]\, dt} \tag{34b}$$

This is the simplest and most direct formulation for calculation of mean rate of flow per gram of a complex tissue. But it applies only when tracer is delivered to a common artery as a near-instantaneous slug, and thereby is delivered to each substituent tissue in proportion to local flow rate (paragraph 15). For example, in Eq. (29b), the coefficients I_a and I_b otherwise would not be in the correct ratio to total dose. (See also paragraph 13 regarding possible correction for recirculation.) Note that the curve of activity which appears in the denominator may be of any form, whether multiexponential or not. All that is required is a measure of subtended area to a point where readings approach zero.

EXAMPLES

18. Assign the following values for the model of Figure 50 when dose is administered as an instantaneous bolus to the main artery,

$$F_a = 2 \text{ ml/min}, \qquad F_b = 8 \text{ ml/min}$$
$$m_a = 9 \text{ gm}, \qquad m_b = 16 \text{ gm}$$
$$\lambda_a = \tfrac{2}{3}, \qquad \lambda_b = \tfrac{5}{4}$$

Equation (22c):

$$V_a = (\tfrac{2}{3})(9) = 6 \text{ ml}, \qquad V_b = (\tfrac{5}{4})(16) = 20 \text{ ml}$$

Equation (22d):

$$k_a = \tfrac{2}{6} = \tfrac{1}{3}, \qquad k_b = \tfrac{8}{20} = \tfrac{4}{10}$$

Equation (26):

$$f_a = (\tfrac{2}{3})(\tfrac{1}{3}) = \tfrac{2}{9} \text{ ml/gm/min}, \qquad f_b = (\tfrac{5}{4})(\tfrac{4}{10}) = \tfrac{1}{2} \text{ ml/gm/min}$$

Equation (27):

$$f_{mean} = (\tfrac{2}{9})(\tfrac{9}{25}) + (\tfrac{1}{2})(\tfrac{16}{25}) = \tfrac{2}{5} \text{ ml/gm/min}$$

Equation (29b):

$$R(t)/R_0 = 0.2e^{-(1/3)t} + 0.8e^{-(4/10)t}$$

Equation (32):

$$f_{\text{mean}} = \frac{1}{(0.2/\frac{2}{9}) + (0.8/\frac{1}{2})} = \frac{2}{5} \text{ ml/gm/min}$$

Weighted mean distribution coefficient:

$$\lambda_{\text{mean}} = (\tfrac{2}{3})(\tfrac{9}{25}) + (\tfrac{5}{4})(\tfrac{16}{25}) = \tfrac{26}{25}$$

Equation (34a):

$$F = \frac{(25)(26/25)}{\int_0^\infty (0.2e^{-(1/3)t} + 0.8e^{-(4/10)t})\, dt}$$

$$= \frac{26}{(0.2/\frac{1}{3}) + (0.8/\frac{4}{10})} = 10 \text{ ml/min}$$

Equation (34b):

$$f_{\text{mean}} = \frac{26/25}{2.6} = \frac{2}{5} \text{ ml/gm/min}$$

CHAPTER 12

VARIOUS APPROXIMATIONS FROM CURVES

Behavior of Primary Labeled Pool

TURNOVER RATE CONSTANT AND RATE

1. Theoretically, the turnover rate constant (k_{aa}) of the primary labeled pool is calculable in very simple fashion without foreknowledge of the number of associated secondary pools, or the number and location of channels of inflow, outflow, and interchange in the system. This is because k_{aa} is not influenced by the configuration of the model. Equation (18a) of Appendix II will confirm that it is a simple function of slopes and intercepts of the curve for pool a in all models. In the very first instant after zero time the only movement of tracer is *out* of pool a. None is returning from other pools because none as yet exists outside pool a. At a very early time pool a is behaving as though it were a single pool with one-way efflux, and k_{aa} is simply the rate constant of efflux. This rate constant is the instantaneous initial slope of an observed curve describing the change either in fraction of dose, amount of dose, or specific activity in the pool. But this initial downslope, being effectively instantaneous, cannot be directly viewed in a curve which, to be seen at all, must extend beyond the first instant. However, although not apparent to inspection, the initial downslope may be evaluated mathematically at the limit $t = 0$. As explained in Chapter 2, paragraph 6, the intercepts of the exponential components of a curve for either SA or quantity of tracer, when normalized as fraction of their respective sum at t_0, become respective coefficients, H, of the equation for the curve. Thus for SA,

$$SA(t)/(SA)_0 = H_1 e^{-g_1 t} + H_2 e^{-g_2 t} + \cdots + H_n e^{-g_n t} \tag{1}$$

The instantaneous slope at any time is the derivative of this function:

$$\frac{d(SA)}{dt} = -H_1 g_1 e^{-g_1 t} - H_2 g_2 e^{-g_2 t} - \cdots - H_n g_n e^{-g_n t}$$

When this is evaluated for instantaneous downslope at t_0, the expression is

$$\text{slope at} \quad t_0 = -H_1 g_1 - H_2 g_2 - \cdots - H_n g_n \tag{2}$$

Except that the signs are negative (indicating a downward slope) this is the same as that for k_{aa} in Eq. (18a) of Appendix II. If pool size, Q_a, is independently known, or is estimated by extrapolation (Chapter 2, paragraph 5), then turnover rate for pool a is $k_{aa} Q_a$. The foregoing approach would appear to be very useful because nothing need be known about the model as a whole. But as explained in Chapter 2, paragraphs 25 and 26, serious obstacles may be encountered. In essence, these are: (1) an uncertain contour of the early portion of a curve, and (2) an uncertainty as to which material(s) can be assigned to the very *first* pool.

INDIVIDUAL RATE CONSTANT OF TRANSFER FROM PRIMARY POOL

2.   The rate constant for transfer to a secondary pool from labeled pool a is theoretically calculable from a time curve for tracer content of the secondary pool (e.g., from its SA curve if pool size is independently known). Knowledge of the model as a whole is unnecessary. The curve for quantity in pool b when normalized to an expression for fraction of dose as a function of time is

$$q_b/q_{a0} = K_1 e^{-g_1 t} + K_2 e^{-g_2 t} + \cdots + K_n e^{-g_n t}$$

(See Chapter 2.) The instantaneous initial upslope of this curve at zero time is determined by the fraction of dose received per unit time from pool a, which is the same as fraction of dose transferred *from* pool a per unit time, which is k_{ba}. Again, it is assumed that at this early time tracer is not returning to pool a. The derivative of the foregoing expression evaluated at $t = 0$ is

$$\left[\frac{d(q_b/q_{a0})}{dt}\right]_{t=0} = -K_1 g_1 - K_2 g_2 - \cdots - K_n g_n \tag{3}$$

This is the same expression for k_{ba} as in Eq. (23a), Appendix II. Negative signs for one or more coefficients, K, will make the net calculated value positive (indicating an initial upslope). Inaccuracies in defining the early portion of the curve via observed data points pose the same problem as just described for assessment of k_{aa}.

Iodide Transport

THE T/S RATIO

3. Altered physiologic function of the thyroid gland has been assessed by tracer methods which give a semiquantitative estimate of rate constants or clearance constants without definitive compartment analysis (Wollman and Reed, 1962; Newcomer, 1967). In the simplified model of Figure 51 the conversion of iodide to hormonal iodine is not shown because it has been blocked by an inhibiting drug. A two-pool iodide system remains. By the trapping action of thyroid cells the iodide ion in thyroid water (pool *b*) is

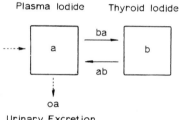

Plasma Iodide   Thyroid Iodide

Urinary Excretion

FIG. 51. A simplified model for iodide transport. Dashed arrows indicate relatively slow external transfer as compared to rates of interchange.

maintained at a much higher concentration than in plasma water (pool *a*). If technically feasible, this gradient would be demonstrable by direct chemical analysis. If the system were *closed*, i.e., if excretory loss from plasma were nonexistent, the exact gradient could be demonstrated in simple fashion by injecting a single dose of tracer iodide into plasma and measuring the ratio of concentration of *tracer* iodide in thyroid water (c_b) to that in plasma or serum water (c_a) at equilibrium. This ratio (T/S ratio) would be directly related to the function of the cells in maintaining the gradient. In other words it would correlate directly with the relative magnitude of the ratio of clearance, Cl_{ba}, to clearance, Cl_{ab}. As noted in Chapter 8, paragraph 7, the following relationship would exist at equilibrium,

$$Cl_{ba}/Cl_{ab} = c_b/c_a = \text{T/S ratio} \tag{4}$$

and the concentration ratio of tracer in water is matched by the same ratio for unlabeled iodide ion in water. A drug, such as thiocyanate, which reduces trapping of iodide by the thyroid, will cause the T/S ratio to fall. In Figure 51 both the input to the system from the outside and the exit (*oa*) are shown as dashed arrows to indicate that these rates are quite small in relation to interchange rates *ba* and *ab*. In this circumstance the system is effectively a closed one, and little error is introduced into Eq. (4) by assuming that Cl_{ba} is the sole exit.

INDIVIDUAL EXCHANGE CONSTANTS

4.   The rate of change of quantity of tracer iodide in thyroid (pool *b*), if expressed in terms of two definitive pools, would be given by the following differential equation,

$$dq_b/dt = q_a k_{ba} - q_b k_{ab} \tag{5}$$

However, the amount of tracer in plasma (q_a) will not be directly measured. In fact, this model is oversimplified, in that plasma iodide equilibrates with a much larger pool of iodide in extracellular fluid. Consequently, it would be helpful to use clearance to thyroid from plasma, i.e., Cl_{ba}. This represents gross clearance without correction for reflux back to plasma. The term $q_a k_{ba}$ is easily alterable. It represents rate of movement of tracer (actually as a function of time, although $q_a(t)$ is written q_a for simplicity). Hence, it is the same as $Cl_{ba} c_a$ when c_a is concentration of tracer in plasma (water). Thus, Eq. (5) may be written

$$dq_b/dt = Cl_{ba} c_a - q_b k_{ab} \tag{6}$$

Dividing all terms by V_b, the volume of water in the thyroid gland, changes q_b to c_b, the concentration of tracer in thyroid water,

$$dc_b/dt = Cl_{ba} c_a/V_b - c_b k_{ab} \tag{7}$$

For a convenient approximation consider that thyroid weight, m_b, (e.g., milligrams) is the same as that of its participating volume, V_b, (e.g., microliters). Then,

$$dc_b/dt = (Cl_{ba} c_a/m_b) - c_b k_{ab} \tag{8}$$

The variable, c_b, likewise may be construed either as units of tracer per milligram or per microliter. One choice of integral for Eq. (8) is

$$c_b(t_2) = c_b(t_1)e^{-k_{ab}(t_2-t_1)} + (Cl_{ba}/m_b)e^{-k_{ab}t_2}\int_{t_1}^{t_2} c_a(t)e^{+k_{ab}t}\, dt \tag{9}$$

Concentration of tracer in the thyroid at t_2 is the left-hand side of the equation, and concentration at prior t_1 is designated as $c_b(t_1)$ in the first term on the right. Remaining unintegrated at the end is an expression which can be plotted with trial values of k_{ab} to give a measurable area between two points in time, t_1 and t_2. Animals taken at successive intervals provide mean thyroid weight and mean values for tracer concentration in thyroid (c_b) along with mean values for tracer concentration in serum (c_a) as a function of time. To be evaluated are the rate constant k_{ab} and the clearance constant Cl_{ba}/m_b (the latter is microliters of blood iodide cleared per unit time *per milligram thyroid tissue*). Solution is by trial and error. Various trial values of the two unknowns are substituted until the equation balances reasonably well for several choices of t_1 and t_2. Start with the expression under the integral sign and construct a curve for $c_a(t)e^{+k_{ab}t}$, with c_a being observed at time t and k_{ab} being a trial value. The area under the curve between t_1 and t_2 is then combined with the remainder of the equation which contains a trial value also for Cl_{ba}. The formula does not require that tracer be delivered rapidly to blood. The injection can be intraperitoneal. Solution of the equation for best fit might best be performed with a digital computer programmed for least squares approximation.

Impulse Analysis

5. The following method of analysis may be less familiar to biomedical scientists than to engineers. Its applicability to a biologic system was first noted by Stephenson (1948), and subsequently discussed by Meier and Zierler (1954), Sheppard (1962), and Bassingthwaighte (1970). The method is classed as stochastic (Chapter 5, paragraphs 1 and 2). Classically, the reference model is the " black box " of Figure 52A. Its effect is to perform a systematic operation on the contents of the system (or on what passes through it). In a hydrodynamic system its interior may consist of an uncharacterized assortment of pools or tubular channels. In a chemical system the box may represent a single compound which yields another compound (represented by output), or a series of intermediates between precursor (input), and product (output). If a time–concentration curve or time–SA curve for tracer is recorded at input the interposed system will remold the curve in accordance with an inherent characteristic which may be called the *transport function* of the enclosed system.

TRANSPORT FUNCTION

Impulse

6. Obviously if one were to characterize the time course of passage of tracer through a system, a sharply defined zero reference point would be required. The tracer should be delivered "instantaneously" at the entrance.

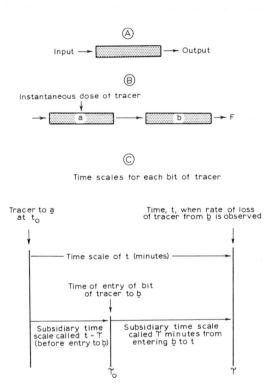

FIG. 52. Models and diagram to illustrate time intervals in the convolution integral.

Such delivery is defined as an *impulse*. It exists only in theory because it would plot as a spike with zero width. In mathematical terms such a spike is called the *delta function*. In ordinary compartment analysis the delivery also is assumed to be instantaneous, although in practice this need only mean that it is rapid in relation to its rate of movement within the system. Such relatively rapid delivery is sometimes called *pulse labeling*. One purpose of the analytic approach which follows is to predict the time–curve for tracer output from a system if a true impulse were delivered at point of natural inflow (Figure 52A). This prediction is possible even if the span of delivery of tracer is quite long, provided SA of tracer versus time (or concentration, as the case may be) is measured in the inflowing donor stream, and also at point of output. What is predicted is fraction of dose lost per unit time at output if an impulse were delivered at zero time. This is known variously as the *transport function, transfer function, impulse response function, admittance function*, or *weighting function* of the system. We will refer to it as the transport function.

Transport function is a time curve

7. As explained in Chapter 7, a concentration curve observed at the point of exit from a system reflects the distribution of transit times for separate bits of the dose. Timing is between point of placement of dose and point of exit. Because the concentration of tracer in efflux varies with time, the amount being lost per unit of time varies at each instant in proportion to its concentration. *If the amount being lost per unit time at each instant is expressed as fraction of dose, this is the transport function of the system.* If F is the constant output rate of tracee, $c(t)$ is tracer concentration in efflux at a given instant, and D is an impulse dose of tracer, then $F \cdot c(t)$ is amount of tracer being lost per unit time at that instant and $F \cdot c(t)/D$ is the corresponding expression for instantaneous rate of loss in terms of fraction of dose. This fractional rate of loss at any point in time will be called h. Being a function of time, it will be written $h(t)$. When plotted versus time, it generates a curve. From the foregoing definition, the equation for transport function may be written,

$$h(t) = Fc(t)/D \tag{10}$$

To convert a time–concentration curve (or one of SA) of effluent, already expressed as fraction of pulse dose per unit volume, to a curve for $h(t)$, simply multiply all points on the concentration curve by output rate. For example, in Figure 13 consider the two pools as jointly comprising a single system. The curves for transport function of the system to match each of the assorted choices of assigned output rates are obtained by multiplying points on each curve by 0.0001, 0.001, etc.

Transport function is probability density

8. Fraction lost per unit time multiplied by a specific time interval gives fraction lost in this interval, however the interval must be very short, i.e., Δt, because rate of loss is continually changing,

$$\text{fraction of dose lost during} \quad \Delta t = h(t)\,\Delta t \tag{11}$$

In the parlance of probability theory $h(t)$ is *probability density*, and the product in Eq. (11) expresses the *probability* that this fraction will be lost during the interval Δt. (In a practical sense it is in fact the fraction lost.) For a protracted time span it is the summation of a sequence of such products (area under the curve between the points delimiting the span),

$$\text{fraction of dose lost in the interval between } t_1 \text{ and } t_2 = \int_{t_1}^{t_2} h(t)\,dt$$

Between zero and infinite time the whole dose is lost. Thus,

$$\int_0^\infty h(t)\,dt = 1 \tag{12}$$

and the probability is unity that the whole dose will eventualiy be lost. In Figures 14 and 16, for example,

$$h(t) = 0.04(-0.26e^{-0.037t} + 0.26e^{-0.0081t})$$

$$\int_0^\infty h(t)\, dt = -(0.0104/0.037) + 0.0104/0.0081 = 1$$

Note that because each point on the curve for $h(t)$ is transit time for that bit of dose, the transport function also may be designated *probability density of transit times*, or *frequency function of transit times*.

Equivalent expressions

9.   Equation (12) obviously assumes that all output from the system is via a single exit. Thus, rearrangement of Eq. (10) gives total flow rate through the system,

$$F = h(t)D/c(t) \tag{13a}$$

Because all input mixes with tracer at the site of entry the Stewart–Hamilton equation also applies,

$$F = D\Big/ \int_0^\infty c(t)\, dt \tag{13b}$$

Equating the right sides of Eqs. (13a) and (13b) produces another expression for $h(t)$,

$$h(t) = c(t)\Big/ \int_0^\infty c(t)\, dt \tag{14}$$

Units of concentration, $c(t)$, in Eq. (14) are not specified. Conversion to other units by multiplying numerator and denominator by a constant is quite permissible. Multiplying $c(t)$ by a conversion factor for a monitoring detector converts $c(t)$ to $R(t)$. Note also that the expression for *mean* transit time for tracer in terms of concentration of tracer at point of sole efflux (Eq. (5), Chapter 7) when multiplied by F/D above and below, changes $c(t)$ to $h(t)$. Then

$$t_{mean} = \int_0^\infty th(t)\, dt \Big/ \int_0^\infty h(t)\, dt$$

The denominator is unity (Eq. (12)). Thus,

$$t_{mean} = \int_0^\infty th(t)\, dt \tag{15}$$

THE CONVOLUTION INTEGRAL

Purpose

10. The integral to be described may be used to obtain a curve for $h(t)$ when, as noted in paragraph 6, the time-curves for concentration of tracer (or SA) at points of input and output of a system are available. In Figure 52B the "black box" on the right (b) represents the uncharacterized system to be evaluated for transport function, $h_b(t)$. The box on the left (a) exists solely for the purpose of generating a time–concentration function to represent input of tracer to b.

Schematic basis for the convolution integral

11. The concept behind the convolution integral is best approached by considering a dose of tracer as a collection of small "bits" each of which has a separate time of transit from one location to another. In Figure 52C the time from t_0 to t is that required for a given bit of the dose placed in a at t_0 to leave b at t. A subsidiary time scale, with units τ, applies to the portion of total time spent in b. The instant the bit enters b is called τ_0, and time of sojourn in b is τ. In terms of the basic scale of t the prior time spent in a can be called $t - \tau$. Although rate of delivery (F) of unlabeled material to b is constant, the rate of delivery of tracer is not constant because its concentration in input varies with time (i.e., as the concentration function generated by a). Call the instantaneous rate of delivery r units per minute, and consider it, rather than a function of t, a function of $t - \tau$. Thus it is written $r(t - \tau)$, which reads "r as a function of $t - \tau$." Upon entrance of tracer into b at rate $r(t - \tau)$, its existence in b is timed on the scale of τ. The *amount* moving which has entered during a short time $\Delta\tau$ is *rate* multiplied by *time*. Thus,

$$\text{amount in bit entering near} \quad \tau_0 \quad \text{during} \quad \Delta\tau = [r(t - \tau)]\,\Delta\tau \quad (16)$$

Having entered b this bit of tracer has an infinite number of possible transit times through b. Thus, it may be considered to be divided into smaller subbits each with a different transit time as predicted by the transport function of b, which is $h_b(\tau)$. The latter is fraction of the total bit lost per unit of time (e.g., per minute) at time τ, which latter is the same as t on the scale of t. Thus, multiplying $h_b(\tau)$ (for fraction of bit lost per minute) by the quantity in the whole bit, as defined by Eq. (16), defines the quantity of that bit of tracer entering at τ_0 lost per minute at time t. Thus,

quantity lost per minute at t represented by this bit

$$= [h_b(\tau)][r(t - \tau)][\Delta\tau] \quad (17)$$

Between t_0 and t there will be a succession of such bits with different time points (τ_0) for entry into b. Adding all their contributions to the rate at t gives overall rate of loss of tracer in units of fraction of dose per minute at t. Such addition is represented as the integral of Eq. (17),

$$\text{total units of tracer lost per minute at time } t = \int_0^t [h_b(\tau)][r(t-\tau)]\, d\tau \quad (18)$$

The integral on the right is a *convolution integral*. The name implies that something is folded (*Faltung* integral). What is folded graphically is one of the curves. For example, the explicit input curve to c generated by the a–b complex in Figure 53 is folded to the left as a mirror image about its zero time ordinate, because this is the orientation "seen" by recipient pool c. Expression of time as $t - \tau$ accomplishes this folding. (See Appendix IV.) The integral as expressed in terms of concentration functions will be given later. The transport function, although written as a function of τ for this integral, is intrinsically the same as $h(t)$. Of interest is the list of names attached to this integral. Although it is sometimes called Duhamel's integral, Borel and others are connected historically with its use. It also is called the *superposition integral* or the *composition product*, and sometimes it is called a *theorem* rather than an integral.

Adaptation of integral to units of concentration

12.   The rate of tracer loss given by Eq. (18) is the same as steady state input–output rate, F, multiplied by the instantaneous concentration of tracer

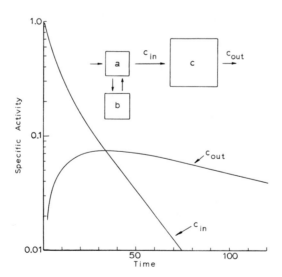

FIG. 53. A model to generate known curves to illustrate deconvolution via an inverse transform.

in output, $c(t)$. Moreover, r can be replaced by the product of F and the tracer concentration in input, $c(t)_{in}$. Then the equation becomes

$$F \cdot c(t)_{out} = \int_0^t [h_b(\tau)][F \cdot c(t - \tau)_{in}] \, d\tau$$

The Fs, being constant, vanish to leave

$$c(t)_{out} = \int_0^t [h_b(\tau)][c(t - \tau)_{in}] \, d\tau \tag{19}$$

This is a working version of the convolution integral. The two variables said to be *convolved* are $h_b(t)$ and $c(t)_{in}$. A shorthand designation for the convolution is $h_b(t) * c(t)_{in}$. The transport function, $h_b(t)$, may be considered an "operator" which, when applied to input, generates output. It also may be construed as a weighting function. It determines, as a function of time, the weight or importance to be assigned what is conceived to be a series of separate inputs of tracer. Equation (19) can also be written with the functions interchanged, i.e.,

$$\int_0^t [h_b(t - \tau)][c(\tau)_{in}] \, d\tau$$

If the concentration curves $c(t)_{in}$ and $c(t)_{out}$ are available the curve for $h(t)$, here $h_b(t)$, emerges by a process of deconvolution to be described. To illustrate the graphical counterpart of convolution, plot the input curve, i, of Figure 54 (see paragraph 17) on thin paper, turn it over (so that abscissal numbering runs from right to left) and overlay it on the curve for $h_c(t)$ of recipient pool c plotted to the same abscissal scale in conventional left to right orientation. The latter curve is shown in Figure 55, and equations are given in paragraph 17. Align the origin for i over the abscissal point of 10 for $h_c(t)$. In the context of convolution the latter now is $h_c(\tau)$. The applicable equation is

$$j = \int_0^t [h_c(\tau)][i(t - \tau)] \, d\tau \tag{19a}$$

The value of t in Eq. (19a) now is 10 and j is an area which may be demonstrated as follows: At $\tau = 1$ on the bottom curve the value for $h_c(\tau)$ is 0.0244 and for overlaid i at this aligned point (9 on its scale) is 0.155. The product of these two is 0.0038. At $\tau = 2$ (where i is 8) the respective values are 0.0238, 0.152; product 0.0036. Successive products for $\tau = 3$, $i = 7$, etc. up to $\tau = 10$, $i = 0$, yield a curve in the interval $t = 0$, $t = 10$, the subtended area of which is the value of the output curve, j, at 10 min. For its value at 15 min move the origin of overlaid i to where $\tau = 15$ and repeat the foregoing process. Successive shifts will give a succession of values for j. (See Figure 56 (in Appendix IV) for illustrative graphics.)

Relation of convolution to Laplace transforms

13.   The mathematical characteristics of Laplace transformation are such that the Laplace transform, $\mathscr{L}$, of the convolution integral is the product of the transforms of the two components in the integrand. Thus,

$$\mathscr{L}\left\{ \int_0^t [h(\tau)][c(t-\tau)_{in}] \, d\tau \right\} = \mathscr{L}\{h(t)\} \cdot \mathscr{L}\{c(t)_{in}\} \tag{20}$$

or

$$\mathscr{L}\{c(t)_{out}\} = \mathscr{L}\{h(t)\} \cdot \mathscr{L}\{c(t)_{in}\} \tag{21}$$

The practical significance of this relationship will be shown when deconvolution is described.

Linearity

14.   The convolution principle is said to apply only to a *linear* system. A steady-state system, as defined in Chapter 10, paragraph 1, is automatically linear. In the parlance of systems analysis in engineering a linear system is one in which the "responses sum." Although "response" may imply that input is a stimulus, the effect of introducing nonperturbing tracer on an output curve of SA is analogous. For illustration take Model A of Figure 19. Call pool *b* the "system" receiving input which will be construed as the SA curve generated by pool *a*. Output will be construed as the SA curve for pool *b* (which is observed in material at the output site). Give two successive pulse doses of 1 unit each,

	Input curve		Output curve
First trial	$1e^{-0.02t}$	$\rightarrow$	$-1e^{-0.02t} + 1e^{-0.01t}$
Second trial	$1e^{-0.02t}$	$\rightarrow$	$-1e^{-0.02t} + 1e^{-0.01t}$
Sum	$2e^{-0.02t}$		$-2e^{-0.02t} + 2e^{-0.01t}$

The sum of the plotted curves will be twice as high *at all points* for both input and output. If the exponential, $e^{-0.01t}$, changed between trials because F_{ob}/Q_b changed, this would not be so. Note also that Eq. (19) is not derivable from the one preceding it if F_{in} does not equal F_{out}. Theoretically, a system could be linear by this definition, though not in steady state, if input–output rates and the sizes of the contained pools all changed in perfectly constant ratio between trials. The assumption of linearity is always invoked for the convolution principle because it assumes that if input of tracer is protracted the output curve is equivalent to the summation of a succession of input pulses (superposition).

DECONVOLUTION

Via inverse transforms

15. Equation (21) may be employed to assess $h(t)$ provided the curves for $c(t)_{in}$ and $c(t)_{out}$ can be expressed in a mathematical form upon which the inverse Laplace transformation may be performed. Laplace transformation is ordinarily used as a mathematical tool to simplify integration. If $f(t)$ is any expression which varies as a function of time, the transform $\mathcal{L}\{f(t)\}$ is the integral of $f(t)$ multiplied by an integrating factor, e^{-pt}. Thus,

$$\mathcal{L}\{f(t)\} = \int_0^\infty e^{-pt} \cdot f(t)\, dt$$

Equation (21), therefore, can be written

$$\int_0^\infty e^{-pt} \cdot c(t)_{out}\, dt = \int_0^\infty e^{-pt} \cdot h(t)\, dt \cdot \int_0^\infty e^{-pt} \cdot c(t)_{in}\, dt \tag{22}$$

The symbol p is a "dummy" variable of the transformation. For illustration, suppose that the curves for $c(t)$ or $SA(t)$ are complex exponentials derived by curve peeling. The curves of Figure 53 have been normalized so that the value of $c(t)_{in}$ at onset is called 1 and all other values on both this curve and the other are *fractions* of this starting value. Coefficients, therefore, are comparable to those designated as H and L in Chapters 2 and 3. The curve for $c(t)_{in}$, in general form, is

$$c(t)_{in} = I_1 e^{-g_1 t} + I_2 e^{-g_2 t} \tag{23a}$$

Specifically, in this example,

$$c(t)_{in} = 0.6e^{-0.2t} + 0.4e^{-0.05t} \tag{23b}$$

The curve for $c(t)_{out}$ in general form is

$$c(t)_{out} = I_1' e^{-g_1 t} + I_2' e^{-g_2 t} + I_3' e^{-g_3 t} \tag{24a}$$

Specifically, in this example,

$$c(t)_{out} = -0.032e^{-0.2t} - 0.1e^{-0.05t} + 0.132e^{-0.01t} \tag{24b}$$

In the model of Figure 53 the a–b complex of pools serves simply to generate a tangible input curve, $c(t)_{in}$, to be accepted by pool c which latter pool is considered to be the "system" for which the transport function, $h_c(t)$, is sought.

16. The right-hand sides of Eqs. (23a) and (24a) are used, respectively, in place of $c(t)_{in}$ and $c(t)_{out}$ in Eq. (22), and all products with e^{-pt} are integrated to obtain expressions in terms of the variable p. The inverse transform restores the variable, t, and eliminates the transformation variable, p. The result is

$$h_c(t) = \frac{(g_1 - g_3)(g_2 - g_3)I_3'}{I_1(g_2 - g_3) + I_2(g_1 - g_3)} e^{-g_3 t} \tag{25}$$

Note that intercepts I', are for output and I are for input. The g appearing in the exponent (here, g_3) represents the particular slope which has been *added* by the recipient pool. Solution for Eq. (25) gives

$$h_c t = 0.01 e^{-0.01t}$$

This is the same expression for $h_c(t)$ as would be obtained via Eq. (10) if pool c were labeled *directly* by instantaneous delivery. In this model $Q_b = 9.09$ and $F_{oc} = 0.0909$. If a dose of 1 were placed *directly* in c, Eq. (10) would give

$$h_c(t) = (0.0909) \frac{1/9.09}{1} e^{-0.01t} = 0.01 e^{-0.01t}$$

Equation (25), of course, was derived for this particular example; however, comparable expressions may be developed for other systems yielding curves which may be expressed as well defined functions and integrated in the transformation process.

Via numerical sequences

17.   A variety of methods for estimating $h(t)$ by deconvolution have been mentioned by Bassingthwaighte (1970). Appendix IV gives an analysis which is applicable to any pair of curves, whether or not their mathematical form is known. Figure 54 happens to be a compartment model with known values. It will serve to check the accuracy of the estimations. The input curve (i) is generated by the two-pool complex ($a + b$). The transport function to be determined is for pool c which has output curve j. The convolution is

$$SA_j(t) = h_c(t) * SA_i(t)$$

Before deconvoluting this expression a prediction may be made for $h_c(t)$ by assuming that pool c is labeled directly and instantaneously with a unit dose. The size of c is 8 mg. Hence,

$$h_c(t) = (0.2)\left(\frac{1/8}{1}\right) e^{-0.025t} = 0.025 e^{-0.025t}$$

The formulas for curves i and j are as follows

Curve i:
$$SA = -\tfrac{1}{3} e^{-0.2t} + \tfrac{1}{3} e^{-0.05t}$$

Curve j:
$$SA = 0.048 e^{-0.2t} - 0.333 e^{-0.05t} + 0.286 e^{-0.025t}$$

A tabulation for n intervals, T, gives the following values for curves i and j where intervals, T (column n), are for 5 units of time:

t	n	i_n	j_n
0	0	0	0
5	1	0.14	0.010
10	2	0.16	0.027
15	3	0.14	0.041
20	4	0.12	0.052

Equations of Appendix IV, paragraph 3, give, for h, at numbered intervals, $n = 0, 1, 2, \ldots$,

$$h_0 = (2)(0.01)/(5)(0.14) = 0.029$$

$$h_1 = [1/(5)(0.14)][0.027 - (5)(0.029)(0.16/2)] = 0.022$$

$$h_2 = [1/(5)(0.14)]\left\{0.041 - 5\left[\frac{(0.029)(0.14)}{2} + (0.022)(0.16)\right]\right\} = 0.019$$

etc. Each calculated value for h is carried to the succeeding equation. The computation is not too onerous with a programmable desk calculator. Figure 55 shows the theoretically correct curve along with points emerging when $T = 5$, as above, and also when $T = 10$. Usually, the smaller the interval for T, the more accurate the estimate (and the more laborious the process). A digital computer is very helpful. The most serious uncertainty in deconvolution occurs in the estimate of early values; however, the errors become "diluted" as these numbers are carried along in the succeeding expressions. The advantage of shortening the first interval used for T may be illustrated in the present example by noting the improved estimate when T is 1 rather than 5 or 10. Then

$$h_0 = (2)(0.00056)/(1)(0.044) = 0.0255$$

A compromise might be to estimate h_0 with $T = 1$, then proceed with $T = 5$ for the curve as a whole. But in an actual working experiment the uncertainty of early data points may be the most serious obstacle to accurate delineation of the early portion of a derived curve. In the present example the simple exponential function for $h(t)$ happens to be easily extrapolable from late well-established points.

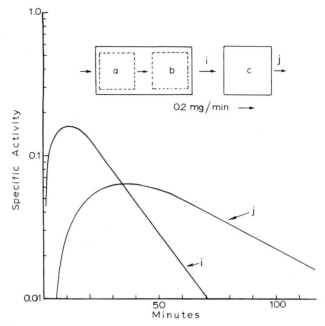

FIG. 54. A model to generate known curves to illustrate deconvolution via numerical sequences.

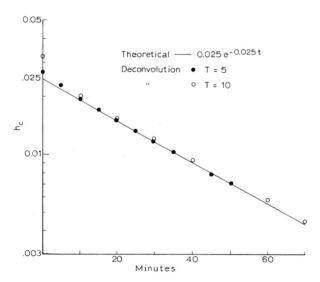

FIG. 55. Theoretical curve for transport function, $h_c(t)$, within pool c of Figure 54. The points represent an approximation to the known curve as obtained by deconvolution through numerical sequences at intervals of either 5 or 10 units of time.

FLUX RATIO IN A REVERSIBLE SYSTEM

18. The model is Figure 32C or 33A. Pools a and b are each labeled on separate occasions or by separate isotopes, and the *amounts* of tracer in a and b versus time are measured. Waterhouse and Keilson (1969) have described the method for a glucose–pyruvate system. The problem is to determine the fraction of a which moves to b, i.e., $F_{ba}/(F_{ba} + F_{oa})$; or the fraction of b which moves to a, i.e., $F_{ab}/(F_{ab} + F_{ob})$. In Model 32C these fractions are $\frac{5}{6}$ and $\frac{1}{3}$, respectively. Discrete pools need not be assumed (e.g., Figure 33A), but it must be possible to define the specific proximate compounds which exchange and obtain time curves for *quantity* of tracer therein. Examine only the avenue F_{ba}. As in paragraph 11 consider that a dose placed in a will move to b and that the delay in reaching b will be measured on the time scale $t - \tau$. As in the case of the classical transport function, the rate of loss of tracer from a in net transfer to b will vary with time and thereby yield a curve comparable to $h(t)$, which describes a succession of transit times. These also may be called *arrival times* at b. Ordinate values will be called S rather than h as previously, because the latter was construed as transport function for an entire unit dose passing through a simple one-way system. In the present model, part of the dose is lost via F_{oa}. The curve for arrival rate is S_{ba} as a function of $t - \tau$, i.e., $S_{ba}(t - \tau)$. The amount of tracer, as *fraction of dose*, which has entered b during a short interval, $\Delta\tau$, near τ_0 will be

$$\text{fraction entered} = [S_{ba}(t - \tau)][\Delta\tau] \tag{26}$$

The next step is to predict what fraction of this bit will exist in b thereafter as a function of time on the scale, τ. This is accomplished by pulse labeling compound b in a separate trial. If dose is *unity*, the time curve as fraction of dose will be $q_b{}^b(\tau)$. It is the fraction of any instantaneously offered tracer (such as that in Eq. (26)) which will exist as a function of time (here, τ). When multiplied by Eq. (26), the product is the time curve for this bit of tracer,

$$\text{fraction of entered bit in } b = [S_{ba}(t - \tau)][\Delta\tau][q_b{}^b(\tau)] \tag{27}$$

For the portion of dose placed in a which is destined to go to b the expected time curve for quantity in b will be a summation of bits, each with its own Eq. (27), i.e., the integral to t of Eq. (27). Experimentally obtain such a curve for fraction of dose in b when label is to a. Call it $q_b{}^a(t)$. From the foregoing development, it should represent such a summation up to any time, t. Thus,

$$q_b{}^a(t) = \int_0^t [S_{ba}(t - \tau)][q_b{}^b(\tau)] \, d\tau \tag{28}$$

This convolution may be written

$$q_b{}^a(t) = S_{ba}(t) * q_b{}^b(t) \tag{29}$$

19.   The curve, $S_{ba}(t)$, for fraction of dose reaching b as a function of time, is the unknown for which the solution is sought by deconvolution. Equation (28) may be expressed as follows in terms of integrals from zero to infinity,

$$\int_0^\infty q_b^a(t)\, dt = \int_0^\infty S_{ba}(t)\, dt \int_0^\infty q_b^b(t)\, dt \tag{30}$$

If the entire dose went to b, rather than partly out F_{oa} as in Figure 32C, the curve for arrival time would be $h_{ba}(t)$ and its integral to infinity would be 1. In the present instance the integral to infinity of the curve $S_{ba}(t)$ obtained by deconvolution represents *that fraction of dose* which goes to b. Unlabeled atoms behaving likewise, it also is the fraction of substance a which is converted to b. Thus,

$$\int_0^\infty S_{ba}(t)\, dt = F_{ba}/(F_{ba} + F_{oa}) \tag{31}$$

This relationship also emerges from the equations for production rate given in Chapter 6, paragraph 3, modified in terms of fraction of dose, q, rather that SA. Label a

$$\text{PR}_b = [F_{ba}/(F_{ba} + F_{oa})]\left[Q_b \middle/ \int_0^\infty q_b^a(t)\, dt\right] \tag{32}$$

Label b,

$$\text{PR}_b = Q_b \middle/ \int_0^\infty q_b^b(t)\, dt \tag{33}$$

Equate Eqs. (32) and (33),

$$\int_0^\infty q_b^a(t)\, dt = [F_{ba}/(F_{ba} + F_{oa})]\left[\int_0^\infty q_b^b(t)\, dt\right] \tag{34}$$

This is convertible to Eq. (30) via Eq. (31). A comparable analysis is possible for fraction of b to a. Label b and sample a; label a and sample a. The convolution is

$$q_a^b(t) = S_{ab}(t) * q_a^a(t) \tag{35}$$

and

$$\int_0^\infty S_{ab}(t)\, dt = F_{ab}/(F_{ab} + F_{ob}) \tag{36}$$

Reversibility, i.e., recycling of label, does not invalidate this calculation. Arrival time is actually *first* arrival time. For example, pools a and b of Figure 32A each may be considered to stand as single pools with two irreversible exits, e.g., pool a discharges to the outside via outlets presently called F_{ba} and

F_{oa}, and pool b via F_{ab} and F_{ob}. The curve for $S_{ba}(t)$ derived by deconvolution will have an equation which is the same as though pool a were separated from b, and then labeled. It would be $SA(t)$ for a multiplied by the present rate F_{ba}. $SA(t)$ would be a simple exponential with slope equal to present $k_{ba} + k_{oa}$. Thus, with unit dose to a,

$$S_{ba}(t) = (1e^{-0.42t})(0.30)$$

Then the integral to infinity of $S_{ba}(t)$ is fraction of a out F_{ba},

$$\int_0^\infty S_{ba}(t)\,dt = 0.30/0.42 = 0.714 = F_{ba}/(F_{ba} + F_{oa})$$

USEFULNESS OF DECONVOLUTION

20. If the purpose of an experiment is to calculate rate of input–output for one-way models such as in Figures 52–54, nothing is gained by deriving transport function via deconvolution. The concentration curve at output gives rate directly via the Stewart–Hamilton equation. Also, this curve gives mean time, i.e., turnover time, and hence the quantity of unlabeled material in the system. One use for transport function as determined by deconvolution is to characterize the general nature of the system. Parrish et al. (1959), for example, concluded that the shape of the curve was consistent with laminar flow throughout the lungs rather than flow through a mixing pool. Bassingthwaighte (1966) drew conclusions concerning the nature of the arterial system in the leg. Waterhouse and Keilson (1969) resorted to deconvolution for analysis of the recycling system just described because certain curves, due to their tendency to flatten out, could not be integrated to infinity. Ordinarily, a model such as that of Figure 32A or 32C is completely solvable by formulas for cumulative excretion such as Eqs. (13) and (14) of Chapter 6, or their counterparts wherein αT is replaced by $\int_0^\infty SA(t)\,dt$ (Eq. (7a) of Chapter 6). Although it does not employ tracer, an interesting application of convolution–deconvolution in nonsteady state has been made by Turner et al. (1971) to assess the rise and fall of insulin production rate after a glucose load.

DERIVATION OF THE FORMULA FOR SIMPLE EXPONENTIAL LOSS

An experimental observation shows that for any given quantity of material present at any point in time a *constant fraction* disappears in a subsequent given interval of time. Thus, the *amount* lost in a given time span is proportional to the amount present at the beginning of the time span. Let q be the amount present at any time, t. The rate of change of q with respect to time is proportional to q, and is negative (decreases),

$$-(dq/dt) \quad \text{is proportional to} \quad q$$

Introduce a proportionality constant (k), i.e., the explicit fraction lost per unit time. Then, the differential equation for fraction of q lost per unit time (rate of decrease) is

$$-(dq/dt) = kq$$

Interchange dt and q, shift signs, and express each side as an indefinite integral,

$$\int dq/q = \int -k \, dt$$

Integration gives

$$\ln q = -kt + \text{constant}$$

The constant of integration is unknown at this point. Call it $\ln C$, and its antilog, C, will ultimately appear in an expression which permits evaluation of C,

$$\ln q = -kt + \ln C$$
$$\ln q - \ln C = -kt$$
$$\ln(q/C) = -kt$$

This says that the product $-kt$ is an exponent of the base e which gives a value equal to q/C,

$$q/C = e^{-kt}$$
$$q = Ce^{-kt}$$

At zero time, the value of e^{-kt} is 1. Thus, at that time $q = C$. The value of C therefore must be q_0, the amount present at zero time,

$$q = q_0 e^{-kt}$$

MATHEMATICAL SOLUTION OF MULTICOMPARTMENT MODELS IN STEADY STATE

A. An Open Three-Pool System

1. The model presented in Figure 20A, Chapter 3, embodies a complete set of entrance channels, exit channels, and interconnections among three pools. Models lacking some of these channels, such as B–K, might be solved separately by setting up the appropriate differential equations for each, and performing separate integrations for each. The advantage in solving the most general version is that integration is accomplished once and for all, and the working equations thereby derived may be adapted algebraically for any more restricted version. The equations in fact are applicable even if the system contains fewer than three pools.

2. The rate of change of quantity of tracer (q) in any pool is established by the composite influence (algebraic summation) of rates of acquisition and rates of loss of tracer. Pool a, which is initially instantaneously labeled at zero time, subsequently receives tracer again as reflux from pool b and pool c. The rate for each of these contributions is represented by the product of the fraction removed from the donor pool per unit time and the quantity of tracer existing in that pool at any instant, i.e., $k_{ab}q_b$ and $k_{ac}q_c$. The individual rates of loss from pool a are $k_{oa}q_a$, $k_{ba}q_a$, and $k_{ca}q_a$, which may be grouped together as a total to be called $k_{aa}q_a$. The differential equation for pool a, therefore, is

Pool a:

$$dq_a/dt = k_{ab}q_b + k_{ac}q_c - k_{aa}q_a \qquad (1)$$

Similarly,

Pool b:

$$dq_b/dt = k_{ba}q_a + k_{bc}q_c - k_{bb}q_b \qquad (2)$$

Pool c:

$$dq_c/dt = k_{ca}q_a + k_{cb}q_b - k_{cc}q_c \qquad (3)$$

3. The problem is to integrate these three simultaneous differential equations so that from the observed time curve for q present in a sampled pool the rate constants of flow (k) and actual flow rates of natural material (F) may be calculated, and the companion curves for the other two pools derived. The functions describing the curves for the three pools as they would be observed, and as will be confirmed by integration of Eqs. (1)–(3), are of the following exponential form when expressed in terms of fraction of dose in a given compartment after the dose (q_{a0}) is introduced into pool a at zero time,

$$q_a/q_{a0} = H_1 e^{-g_1 t} + H_2 e^{-g_2 t} + H_3 e^{-g_3 t} \qquad (4)$$

$$q_b/q_{a0} = K_1 e^{-g_1 t} + K_2 e^{-g_2 t} + K_3 e^{-g_3 t} \qquad (5)$$

$$q_c/q_{a0} = L_1 e^{-g_1 t} + L_2 e^{-g_2 t} + L_3 e^{-g_3 t} \qquad (6)$$

It may be predicted that, as a result of integration, the coefficients H_i, K_i, and L_i $(i = 1, 2, 3)$ will emerge as equivalents of complex expressions involving rate constants, k, and the exponential slopes, g, and that other expressions for rate constants in terms of slopes also will evolve.

4. Integration of Eqs. (1)–(3) may be accomplished through their Laplace transformation

$$\mathcal{L}\left\{\frac{dq_i}{dt}\right\} = \int_0^\infty e^{-pt}\left(\frac{dq_i}{dt}\right) dt = p\mathcal{L}\{q_i(t)\} - q_{i0}$$

where p is a variable introduced as an integrating factor. Let the transform $\mathcal{L}\{q_i(t)\}$ be called x for $q_a(t)$, y for $q_b(t)$, and z for $q_c(t)$. Equations (1)–(3) then become

$$px - q_{a0} = -k_{aa}x + k_{ab}y + k_{ac}z \qquad (7)$$

$$py - 0 = k_{ba}x - k_{bb}y + k_{bc}z \qquad (8)$$

$$pz - 0 = k_{ca}x + k_{cb}y - k_{cc}z \qquad (9)$$

These three simultaneous equations may be solved to give

$$x = \frac{q_{a0}[(p + k_{bb})(p + k_{cc}) - k_{bc}k_{cb}]}{U} \tag{10}$$

$$y = \frac{q_{a0}[k_{ba}(p + k_{cc}) + k_{ba}k_{ca}]}{U} \tag{11}$$

$$z = \frac{q_{a0}[k_{ca}(p + k_{bb}) + k_{ba}k_{cb}]}{U} \tag{12}$$

$$U = p^3 + p^2C_1 + pC_2 + C_3$$

In this symmetrical third-order equation, the values of C_1, C_2, and C_3 may be represented by a sum, summed pairs, and a product of three dissimilar constants g_1, g_2, and g_3. These three combinations of the three constants are made identical with the corresponding series of values for C_1, C_2, and C_3 derived for the denominator of Eqs. (10)–(12) from the algebraic solution of Eqs. (7)–(9); thus,

$$g_1 + g_2 + g_3 = k_{aa} + k_{bb} + k_{cc} \tag{13}$$

$$g_1g_2 + g_1g_3 + g_2g_3 = k_{aa}k_{bb} + k_{aa}k_{cc} + k_{bb}k_{cc}$$
$$- k_{ab}k_{ba} - k_{bc}k_{cb} - k_{ac}k_{ca} \tag{14}$$

$$g_1g_2g_3 = k_{aa}k_{bb}k_{cc} - k_{aa}k_{bc}k_{cb} - k_{bb}k_{ac}k_{ca}$$
$$- k_{cc}k_{ab}k_{ba} - k_{ab}k_{bc}k_{ca} - k_{ac}k_{ba}k_{cb} \tag{15}$$

5. Proceeding hence by an inverse Laplace transform† the equation for pool a becomes that of Eq. (4), where the constants g_1, g_2, and g_3 now are identified with the exponential slopes g_1, g_2, and g_3, and

$$H_1 = \frac{(-g_1 + k_{bb})(-g_1 + k_{cc}) - k_{bc}k_{cb}}{(-g_1 + g_2)(-g_1 + g_3)} \tag{16}$$

$$H_2 = \frac{(-g_2 + k_{bb})(-g_2 + k_{cc}) - k_{bc}k_{cb}}{(-g_2 + g_3)(-g_2 + g_1)} \tag{17}$$

$$H_3 = \frac{(-g_3 + k_{bb})(-g_3 + k_{cc}) - k_{bc}k_{cb}}{(-g_3 + g_1)(-g_3 + g_2)} \tag{18}$$

Only two of the foregoing equations are independent, because $H_1 + H_2 + H_3 = 1$. Any two of the above, along with Eqs. (13)–(15), are used jointly to

† "Handbook of Mathematical Tables" (2nd Ed.), Eq. (19), p. 397. Chemical Rubber Co., Cleveland, Ohio, 1964.

solve for rate constants. A convenient initial working equation emerges from a combination of any two of Eqs. (16)–(18) along with Eq. (13) and the fact that the values of H add to 1. It is

$$k_{aa} = H_1 g_1 + H_2 g_2 + H_3 g_3 \tag{18a}$$

This equation applies to all models regardless of their configuration.

6. In Eqs. (5) and (6), the values of g_1, g_2, and g_3 are the same as those for Eq. (4), and the following general expressions define values for the coefficients K_i and L_i of the curves for pool b and pool c, respectively. K_i may be K_1, K_2, or K_3, and L_i may be L_1, L_2, or L_3. When the equation is directed toward K_1 the slope is g_1 and the denominator U_1 (Eq. (21)). Similarly, g_2 and U_2 accompany K_2 or L_2, etc.,

$$K_i = \frac{(-g_i + k_{cc})k_{ba} + k_{bc}k_{ca}}{U_i} \tag{19}$$

$$L_i = \frac{(-g_i + k_{bb})k_{ca} + k_{ba}k_{cb}}{U_i} \tag{20}$$

where (as in Eqs. (16)–(18))

$$U_1 = (-g_1 + g_2)(-g_1 + g_3) \tag{21}$$

$$U_2 = (-g_2 + g_3)(-g_2 + g_1) \tag{22}$$

$$U_3 = (-g_3 + g_1)(-g_3 + g_2) \tag{23}$$

Equations (19) and (20), with Eqs. (5) and (6), not only define the curves for the amount of tracer in pools b and c when observations are made in pool a, they also can be used in the solution for unknown rate constants when K_i and/or L_i are obtained by direct sampling of pools b and/or c. Note that, as with H_i, only two of the three equations for K_i and L_i are independent, since $\sum K_i = \sum L_i = 0$. Convenient working equations are obtained by combining Eqs. (19) or (20) with Eq. (13),

$$k_{ba} = -K_1 g_1 - K_2 g_2 - K_3 g_3 \tag{23a}$$

$$k_{ca} = -L_1 g_1 - L_2 g_2 - L_3 g_3 \tag{23b}$$

B. An Open Four-Pool System

7. The foregoing mathematical operations are possible with a model consisting of any number of pools, but beyond three the task of calculation becomes increasingly more formidable. For each additional pool another slope and coefficient are added. In solving for the rate constants from observations in the first pool there are two additional equations and an accompanying increase in complexity of all equations. The working formulations for an interconnected four-pool model $(a, b, c, \text{and } d)$ will illustrate this point,

$$g_1 + g_2 + g_3 + g_4 = k_{aa} + k_{bb} + k_{cc} + k_{dd} \tag{24}$$

$$
\begin{aligned}
g_1 g_2 + g_1 g_3 + g_1 g_4 + g_2 g_3 + g_2 g_4 + g_3 g_4 &= k_{aa}k_{bb} + k_{aa}k_{cc} + k_{aa}k_{dd} \\
&\quad + k_{bb}k_{cc} + k_{bb}k_{dd} + k_{cc}k_{dd} \\
&\quad - (k_{ab}k_{ba} + k_{ac}k_{ca} + k_{ad}k_{da} \\
&\quad + k_{bd}k_{db} + k_{bc}k_{cb} + k_{cd}k_{dc})
\end{aligned}
\tag{25}
$$

$$
\begin{aligned}
g_1 g_2 g_3 &+ g_1 g_2 g_4 + g_1 g_3 g_4 + g_2 g_3 g_4 \\
&= k_{aa}k_{bb}k_{cc} + k_{aa}k_{bb}k_{dd} + k_{aa}k_{cc}k_{dd} + k_{bb}k_{cc}k_{dd} \\
&\quad - (k_{ab}k_{ba}k_{cc} + k_{ab}k_{ba}k_{dd} + k_{ac}k_{ca}k_{bb} + k_{ac}k_{ca}k_{dd} + k_{ad}k_{da}k_{bb} \\
&\quad + k_{ad}k_{da}k_{cc} + k_{aa}k_{bc}k_{cb} + k_{aa}k_{bd}k_{db} + k_{aa}k_{cd}k_{dc} + k_{bb}k_{cd}k_{dc} \\
&\quad + k_{cc}k_{bd}k_{db} + k_{dd}k_{bc}k_{cb} + k_{ab}k_{bc}k_{ca} + k_{ab}k_{bd}k_{da} + k_{ac}k_{ba}k_{cb} \\
&\quad + k_{ac}k_{cd}k_{da} + k_{ad}k_{ba}k_{db} + k_{ad}k_{ca}k_{dc} + k_{bc}k_{cd}k_{db} + k_{bd}k_{cb}k_{dc})
\end{aligned}
\tag{26}
$$

$$
\begin{aligned}
g_1 g_2 g_3 g_4 &= k_{aa}k_{bb}k_{cc}k_{dd} + k_{ba}k_{ab}k_{cd}k_{dc} + k_{ac}k_{ca}k_{bd}k_{db} + k_{ad}k_{da}k_{bc}k_{cb} \\
&\quad - (k_{aa}k_{bc}k_{cd}k_{db} + k_{aa}k_{bd}k_{cb}k_{dc} + k_{aa}k_{bd}k_{cc}k_{db} + k_{aa}k_{bb}k_{cd}k_{dc} \\
&\quad + k_{aa}k_{bc}k_{cb}k_{dd} + k_{ab}k_{ba}k_{cc}k_{dd} + k_{ac}k_{ba}k_{cd}k_{db} + k_{ac}k_{ba}k_{cb}k_{dd} \\
&\quad + k_{ad}k_{ba}k_{cb}k_{dc} + k_{ad}k_{ba}k_{cc}k_{db} + k_{ab}k_{bc}k_{ca}k_{dd} + k_{ab}k_{bd}k_{ca}k_{dc} \\
&\quad + k_{ab}k_{bc}k_{cd}k_{da} + k_{ab}k_{bd}k_{cc}k_{da} + k_{ac}k_{bb}k_{ca}k_{dd} + k_{ac}k_{bd}k_{cb}k_{da} \\
&\quad + k_{ac}k_{bb}k_{cd}k_{da} + k_{ad}k_{bb}k_{ca}k_{dc} + k_{ad}k_{bc}k_{ca}k_{db} + k_{ad}k_{bb}k_{cc}k_{da})
\end{aligned}
\tag{27}
$$

$$
\begin{aligned}
H_i = [&(-g_i + k_{bb})(-g_i + k_{cc})(-g_i + k_{dd}) - (-g_i + k_{bb})k_{cd}k_{dc} \\
&- (-g_i + k_{cc})k_{bd}k_{db} - (-g_i + k_{dd})k_{bc}k_{cb} \\
&- k_{bc}k_{cd}k_{db} - k_{ba}k_{cb}k_{dc}]/U_i
\end{aligned}
\tag{28}
$$

$$
\begin{aligned}
K_i = [&(-g_i + k_{cc})(-g_i + k_{dd})k_{ba} + (-g_i + k_{cc})k_{bd}k_{da} \\
&+ (-g_i + k_{dd})k_{bc}k_{ca} + k_{bc}k_{cd}k_{da} + k_{bd}k_{ca}k_{dc} - k_{ba}k_{cd}k_{dc}]/U_i
\end{aligned}
\tag{29}
$$

$$
\begin{aligned}
L_i = [&(-g_i + k_{bb})(-g_i + k_{dd})k_{ca} + (-g_i + k_{bb})k_{ca}k_{da} \\
&+ (-g_i + k_{dd})k_{ba}k_{cb} + k_{ba}k_{cd}k_{db} + k_{bd}k_{cb}k_{da} - k_{bd}k_{ca}k_{db}]/U_i
\end{aligned}
\tag{30}
$$

$$M_i = [(-g_i + k_{bb})(-g_i + k_{cc})k_{da} + (-g_i + k_{bb})k_{ca}k_{dc}$$
$$+ (-g_i + k_{cc})k_{ba}k_{db} + k_{ba}k_{cb}k_{dc} + k_{bc}k_{ca}k_{db} - k_{bc}k_{cb}k_{da}]/U_i$$

$$(31)$$

$$U_1 = (-g_1 + g_2)(-g_1 + g_3)(-g_1 + g_4) \tag{32}$$

$$U_2 = (-g_2 + g_1)(-g_2 + g_3)(-g_2 + g_4) \tag{33}$$

$$U_3 = (-g_3 + g_1)(-g_3 + g_2)(-g_3 + g_4) \tag{34}$$

$$U_4 = (-g_4 + g_1)(-g_4 + g_2)(-g_4 + g_3) \tag{35}$$

where M_i represents the series of four coefficients for the fourth pool. Only three of the four equations emerging from each expression in Eqs. (28)–(31) are independent, since $\sum H_i = 1$, and $\sum K_i = \sum L_i = \sum M_i = 0$.

C. Change in Equations When Channels or Pools Are Eliminated

OPEN MODELS

8. In practice, a model with a given number of pools will not necessarily have all possible channels of entrance, exit, or interflow. In fact, an explicit solution is not possible unless the number of parameters to be evaluated does not exceed the number of independent equations. When a channel of exit or of interchange is non-existent, its corresponding rate constant is made zero in the various working equations which express the slopes and intercepts as functions of rate constants. This is illustrated in Chapters 2, 3, and 4. Some of the terms in the equations are thereby eliminated. A striking example of such attenuation is Eq. (27) as applied to a four-pool, in-line system (Figure 25A, Chapter 4). Nineteen of the terms disappear, leaving only five to use in the calculation.

9. The "algebraic degradation" which accompanies the progressive elimination of pools is of interest. Starting with a four-pool system, then eliminating one pool, the slope g_4 becomes nonexistent (zero) as do all rate constants with a subscript d. Substitution of these zero values in the equation for H_1 (Eq. 28) (denominator being Eq. (32)) reduces that equation to Eq. (16) for a three-pool model. Expressions for H_2 and H_3 are similarly affected, whereas H_4 becomes zero. Removing still another pool to leave a two-pool system will eliminate g_3 and rate constants with subscript c, and will render all of their multiples zero. Equation (16) then becomes

$$H_1 = \frac{(-g_1 + k_{bb})(-g_1)}{(-g_1 + g_2)(-g_1)} = \frac{g_1 - k_{bb}}{g_1 - g_2} \tag{36}$$

Equation (17) is similarly altered, and Eq. (18) disappears (i.e., $H_3 = 0$). For a one-pool system, nothing but H_1 and g_1 remain. Thus, Eq. (36) becomes $H_1 = 1$, and all equations embodying various combinations of slope values (Eqs. (24)-(27)) reduce to one expression: $g_1 = k_{aa} = k_{oa}$.

Closed Models

10. An interchanging system with no external channels (and therefore no net flow through the system) embodies a special restriction. The curves for quantity in all pools will approach horizontal asymptotes, i.e., their slopes ultimately become zero. For a three-pool system the curve for pool a is

$$q_a/q_{ao} = H_1 e^{-g_1 t} + H_2 e^{-g_2 t} + H_3$$

One of the three slopes (g_3, by the convention adopted that it is the least of the three) may be shown to be zero, and H_3 represents the ultimate asymptotic value of the function, because $e^{-g_3 t}$ becomes 1. The residue of working Eqs. (13)-(23) still applies when g_3 is made zero. These equations in such altered form can be derived *de novo* from differential equations similar to Eqs. (1)-(3) but with boundary conditions stipulating that $q_a + q_b + q_c = q_{a0}$, replacing the original condition that at $t = \infty$, $q_a = q_b = q_c = 0$.

D. Special Case, One-Way Flow, Equal-Sized Pools

11. The formulations of Sections A–C are dependent on the assumption that no two slopes are equal. In the event that two are equal the denominators given by two of the Eqs. (21)-(23) will become zero and the intercepts infinite. Actually, in curve analysis an identity between slopes is not detectable. Yet in a nonreversible system such as Figure 19E it is apparent that as the two pools approach equal size (Figure 19F) the value of the rate constant k_{ba} which is slope g_1 approaches k_{ob} which is slope g_2. The equation for q_b takes an entirely different form in such a model. In the following development the common rate constant–slope will be called k. For a two-pool model

$$dq_b/dt = kq_a - kq_b = kq_{a0}e^{-kt} - kq_b$$

$$q_b = e^{-kt}\left[\int e^{+kt}kq_{a0}e^{-kt}\,dt + C\right] \tag{37}$$

$$q_b/q_{a0} = kte^{-kt}$$

For a series of indefinite length, consider the nth pool beyond the first,

$$\frac{dq_n}{dt} = kq_{(n-1)} - kq_n$$

$$q_n = e^{-kt}\left[\int \frac{e^{+kt}kq_{a0}k^{(n-2)}t^{(n-2)}}{(n-2)!} e^{-kt}\,dt + C\right] \tag{38}$$

$$\frac{q_n}{q_{a0}} = \frac{k^{(n-1)}t^{(n-1)}}{(n-1)!} e^{-kt}$$

For integration, the moiety $k^{(n-1)}/(n-1)!$ is the constant, C, of Eq. (11), Appendix V, and $t^{(n-1)}$ is t^N of that equation. Then,

$$\int_0^\infty (q_n/q_{a0})\,dt = 1/k \tag{39}$$

The value is the same for all pools. For mean time of sojourn of tracer, in the system through the nth pool the integrand of the denominator of Eq. (6), Chapter 7, is Eq. (38) (above), and that of the numerator is Eq. (38) multiplied by t (which changes $t^{(n-1)}$ to t^n). The result is

$$t_{mean} = n/k \tag{40}$$

For timing of the peak of the curve from the nth pool (t_{max}), Eq. (12) of Appendix V, with t^N being $t^{(n-1)}$, becomes

$$dq_n/dt = C(n-1)t^{(n-2)}e^{-kt} - Ckt^{(n-1)}e^{-kt} \tag{41}$$

Equating this to zero gives the time of zero slope (t_{max}),

$$t_{max} = (n-1)/k \tag{42}$$

RELATION OF A RATE RATIO TO VERTICAL SHIFT IN $t_{\max}$

The following is proof that in a one-way two-pool system having input to pool b from the outside (Figures 19B and D) the ratio of SA of pool b at the peak of its curve ($t_{\max}$) to the SA of pool a at this point in time is equal to the ratio F_{ba}/F_{bb}. The equation for SA_a may be written

$$SA_a = \frac{q_{a0}}{Q_a} e^{-k_{aa}t} \tag{1}$$

and that for pool b (see Chapter 2, Eq. (12))

$$SA_b = \frac{q_{a0}}{Q_b} \frac{k_{ba}}{(k_{bb} - k_{aa})} (e^{-k_{aa}t} - e^{-k_{bb}t}) \tag{2}$$

The time for $t_{\max}$ of SA_b is that at which the derivative of Eq. (2) has a value equal to zero (zero slope). The resulting expression is

$$k_{aa} e^{-k_{aa}t_{\max}} = k_{bb} e^{-k_{bb}t_{\max}}$$

or

$$e^{-k_{bb}t_{\max}} = (k_{aa}/k_{bb}) e^{-k_{aa}t_{\max}} \tag{3}$$

Substitution of the right-hand side of Eq. (3) for $e^{-k_{bb}t}$ in Eq. (2) evaluated at $t = t_{max}$ gives

$$(SA_b)_{t_{max}} = \frac{q_{a0}}{Q_b} \frac{k_{ba}}{(k_{bb} - k_{aa})} \frac{(k_{bb} - k_{aa})e^{-k_{aa}t_{max}}}{k_{bb}} \qquad (4)$$

The ratio of Eq. (4) to Eq. (1) at t_{max} is

$$\left(\frac{SA_b}{SA_a}\right)_{t_{max}} = \frac{k_{ba}Q_a}{k_{bb}Q_b} = \frac{F_{ba}}{F_{bb}} \qquad (5)$$

Note

The general formula which predicts the time for t_{max} is derived as follows from the observed curve (where the two exponential slopes are, respectively, called g_1 and g_2 in place of k_{aa} and k_{bb} in Eq. (2)). With these symbols substituted in Eq. (3) and the natural log taken of both sides,

$$\ln \frac{g_1}{g_2} - g_1 t_{max} = -g_2 t_{max}$$

$$t_{max} = \frac{\ln(g_1/g_2)}{(g_1 - g_2)} \qquad (6)$$

DECONVOLUTION BY NUMERICAL SEQUENCE APPROXIMATION

1. A solution of the convolution integral may be approximated by application of the trapezoid rule (Chapter 5, paragraph 6). The method has been discussed by Aseltine (1958), and by Cuénod and Durling (1969). Figure 56 will be used for illustration. Shown in part A is an input curve, $i(t)$, similar to that of Figure 53. Such a curve, as it is "seen" to enter by the recipient system, appears folded to the left as a mirror image. It is plotted this way in Figure 56B along with a conventional plot of the transport function, $h(t)$, shown as a dashed line. The two interact to yield an output function of the system, $j(t)$. In parts B, D, and E of the figure the input curve is advanced to the right at successive equal stages, which, in this example, are 5 units of time, t, each. These time spans on the abscissa, to be called T, will represent the bases of successive trapezoids. At the stage shown in B the overlap of the curves covers one unit of T. Before constructing a trapezoid on this base the nature of its two sides must be understood. The left side has an ordinate value which is the *product* of the value of h on the lower scale where $t = 0$ (*zero* units of T), and the value of i as read from the upper scale where $t = 5$ (*one* unit of T). This product, representing the altitude of the left side of the trapezoid, will be called $h_0 i_1$. Similarly, the height of the right side of the trapezoid will be $h_1 i_0$. Figure 56C shows such a trapezoid. The dashed curve

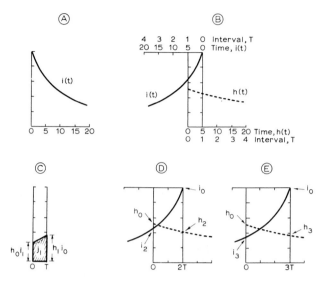

FIG. 56. Graphical representation of the convolution integral. The area enclosed in the trapezoid of part C (bold lines) approximates j_1, the value observed on the output curve at 5 min. The sides of the trapezoid are products of ordinate values for h and i at elapsed times indicated by their respective scales for T in part B.

across the top indicates the true line which would be formed had very closely spaced multiples of aligned values of h and i been used. The straight line across the top is an approximation which improves in accuracy as the chosen interval, T, is made smaller. The present choice of 5 was arbitrary. Equation (19) of Chapter 12, in effect, says that the value of the output curve, $j(t)$, at $t = 5$, is approximately equal to the area of the trapezoid as described. The symbol for j at this point (*one* unit of T) will be j_1 Thus

$$j_1 \approx T\left(\frac{h_0 i_1 + h_1 i_0}{2}\right)$$

2. Next consider part D of Figure 56. The curve for $i(t)$ now is advanced one additional interval of T to the right. The area of overlap is now to be approximated by the sum of two trapezoids, i.e., one in the interval from 0 to T and another from T to $2T$. This total area approximates the value on the curve $j(t)$ at $2T$, i.e., for j_2 (where $t = 10$),

$$j_2 \approx T\left(\frac{h_0 i_2 + h_1 i_1}{2}\right) + T\left(\frac{h_1 i_1 + h_2 i_0}{2}\right)$$

or

$$j_2 \approx T\left(\frac{h_0 i_2}{2} + h_1 i_1 + \frac{h_2 i_0}{2}\right)$$

In the more advanced position shown in part E of the figure, the enclosed area is approximated as the sum of three trapezoids,

$$j_3 \approx T\left(\frac{h_0 i_3 + h_1 i_2}{2}\right) + T\left(\frac{h_1 i_2 + h_2 i_1}{2}\right) + T\left(\frac{h_2 i_1 + h_3 i_0}{2}\right)$$

or

$$j_3 \approx T\left(\frac{h_0 i_3}{2} + h_1 i_2 + h_2 i_1 + \frac{h_3 i_0}{2}\right)$$

For successive values of j corresponding to successive increments of T numbering n,

$$j_n \approx T\left(\frac{h_0 i_n}{2} + h_1 i_{n-1} + h_2 i_{n-2} + \cdots + h_{n-1} i_1 + \frac{h_n i_0}{2}\right)$$

Not included in the foregoing series is the first term, that defining j_0, when the onset of the input curve, $i(t)$, is exactly aligned with the point of onset of the transport function, $h(t)$. The limit, j_0, may be construed as an "empty trapezoid" of zero width and "right-hand" altitude of $h_0 i_0$. The expression would be

$$j_0 \approx T\left(\frac{h_0 i_0}{2}\right)$$

With the sole exception of a single instantaneously mixed pool receiving input as an impulse at zero time, the initial value of j_0 for all systems is zero. Hence,

$$j_0 \approx T\left(\frac{h_0 i_0}{2}\right) = 0$$

However, this expression for j_0 is of no practical use to estimate h_0 since the the result always will be $h_0 = 0$. Yet, as will be pointed out later, h_0 will be greater than zero in a unique receiving system consisting of an instantaneously mixed single pool. Thus the series of expressions which will be given to calculate a sequence of values of h are algebraic rearrangements of the equations for $j_1, j_2, \ldots, j_n$. The observed values are for j and i.

3.   Note that the expression for j_1 contains two unknown values, h_0 and h_1. In a system such as that of Figure 54, the input curve is known to be zero at

$t = 0$. Thus $i_0 = 0$. But the recipient "system" is an instantaneously mixed single pool. Hence h_0 is not zero. The expression for j_1 thus simplifies to

$$j_1 \approx T\left(\frac{h_0 i_1}{2}\right)$$

In terms of h, this expression, and the ensuing ones become

$$h_0 \approx \frac{2j_1}{Ti_1}$$

$$h_1 \approx \frac{1}{Ti_1}\left[j_2 - T\left(\frac{h_0 i_2}{2}\right)\right]$$

$$h_2 \approx \frac{1}{Ti_1}\left[j_3 - T\left(\frac{h_0 i_3}{2} + h_1 i_2\right)\right]$$

$$h_3 \approx \frac{1}{Ti_1}\left[j_4 - T\left(\frac{h_0 i_4}{2} + h_1 i_3 + h_2 i_2\right)\right]$$

$$h_4 \approx \frac{1}{Ti_1}\left[j_5 - T\left(\frac{h_0 i_5}{2} + h_1 i_4 + h_2 i_3 + h_3 i_2\right)\right]$$

$$\vdots$$

$$h_n \approx \frac{1}{Ti_1}\left[j_{n+1} - T\left(\frac{h_0 i_{n+1}}{2} + h_1 i_n + h_2 i_{n-1} + \cdots + h_{n-1} i_2\right)\right]$$

Beginning with initially calculated h_0, values of h are substituted successively in the series of expressions to give a solution for the next value of h.

4. If neither the initial value of the input curve nor of h is zero at $t = 0$ (e.g., Figure 53), the expression for j_1 in paragraph 1, having the two unknowns, h_0 and h_1, is indeterminate. In such a circumstance the function $h(t)$ might be predictable if, as in this instance, the terminal *single* pool dominates the slope of the tail of the output curve. In this example it is 0.01. Therefore $h(t) = 0.01e^{-0.01t}$. However a general approach is to assume that if T is initially made *very short*, the values of h_0 and h_1 are nearly the same. The values i_0 and i_1 likewise might be considered the same, however, both are independently known and may be retained to give an effective mean for the two. Thus, in the expression for j_1 in paragraph 1 substitute h_0 for h_1. Then

$$h_0 \approx \frac{2j_1}{T(i_0 + i_1)}$$

Having made this estimate of h_0 the expressions for the series in terms of j are arranged as follows,

$$h_1 \approx \frac{2}{Ti_0}\left[j_1 - \frac{Th_0 i_1}{2}\right]$$

$$h_2 \approx \frac{2}{Ti_0}\left[j_2 - T\left(\frac{h_0 i_2}{2} + h_1 i_1\right)\right]$$

$$h_3 \approx \frac{2}{Ti_0}\left[j_3 - T\left(\frac{h_0 i_3}{2} + h_1 i_2 + h_2 i_1\right)\right]$$

$$h_4 \approx \frac{2}{Ti_0}\left[j_4 - T\left(\frac{h_0 i_4}{2} + h_1 i_3 + h_2 i_2 + h_3 i_1\right)\right]$$

$$\vdots$$

$$h_n \approx \frac{2}{Ti_0}\left[j_n - T\left(\frac{h_0 i_n}{2} + h_1 i_{n-1} + h_2 i_{n-2} + \cdots + h_{n-1} i_1\right)\right]$$

5. When the recipient system is a complex one wherein entering tracer cannot reach exit instantly at zero time, the transport function is zero at zero time. Then $h_0 = 0$. Rearrangement of the equations for j then yields the following (if i_0 is not zero),

$$h_0 = 0$$

$$h_1 \approx \frac{2j_1}{Ti_0}$$

$$h_2 \approx \frac{2}{Ti_0}\left[j_2 - Th_1 i_1\right]$$

$$h_3 \approx \frac{2}{Ti_0}\left[j_3 - T(h_1 i_2 + h_2 i_1)\right]$$

$$h_4 \approx \frac{2}{Ti_0}\left[j_4 - T(h_1 i_3 + h_2 i_2 + h_3 i_1)\right]$$

$$\vdots$$

$$h_n \approx \frac{2}{Ti_0}\left[j_n - T(h_1 i_{n-1} + h_2 i_{n-2} + \cdots + h_{n-1} i_1)\right]$$

6. If the input curve, $i(t)$, and the transport function, $h(t)$, both begin at zero, the first usable equation is for j_2 in paragraph 2. Making $h_0 = 0$ and $i_0 = 0$ and rearranging this and succeeding expressions for j in terms of h,

$$h_0 = 0$$

$$h_1 \approx \frac{j_2}{Ti_1}$$

$$h_2 \approx \frac{1}{Ti_1}(j_3 - Th_1 i_2)$$

$$h_3 \approx \frac{1}{Ti_1}[j_4 - T(h_1 i_3 + h_2 i_2)]$$

$$h_4 \approx \frac{1}{Ti_1}[j_5 - T(h_1 i_4 + h_2 i_3 + h_3 i_2)]$$

$$\vdots$$

$$h_n \approx \frac{1}{Ti_1}[j_{n+1} - T(h_1 i_n + h_2 i_{n-1} + h_3 i_{n-2} + \cdots + h_{n-1} i_2)]$$

7. Depending on the characteristics of the participating curves, the accuracy of the derived curve for $h(t)$ is quite variable. In general, the estimate improves as the interval, T, is made shorter. It should be noted that an arithmetic feature of the series of equations for h tends to magnify observational errors, particularly in the case of the output curve, $j(t)$. Each computed value for h evolves from a small difference between two larger numbers located in the brackets. Values for j are notably uncertain in a steep portion of its curve smoothed from scattered data points. Estimation of the initial value for h, i.e., h_0, may be particularly poor because its potential parameter, j_1, is not well approximated by the trapezoid geometry. In fact, the trapezoid representation degenerates to that of a triangle when i_0 is zero. However, even if h_0 is made zero in the formulas when the system is such that h should be highest at $t = 0$ (Figure 53), the errors in points h_1, h_2, and a few points thereafter become progressively "diluted" by the increasing number of terms in the formulas which define subsequent values of h. The early approximations tend to oscillate rather widely, perhaps even yielding alternate negative values. Nevertheless, the placement of the early segment of the curve for $h(t)$ may be accomplished rather satisfactorily by graphical averaging of such an array on a linear plot.

SOME DEFINITE INTEGRALS AND DERIVATIVES

Symbols

y Dependent variable (a function of t, i.e., $y(t)$).
C Constant (not in the exponent).
g Constant (in the exponent). Exponential slope.
t Time, as the independent variable.
N Any integer.

If $y = C$,

$$\int_{t_1}^{t_2} C \, dt = C(t_2 - t_1) \tag{1}$$

$$dy/dt = 0 \tag{2}$$

If $y = Ct$,

$$\int_{t_1}^{t_2} Ct \, dt = C(t_2{}^2 - t_1{}^2)/2 \tag{3}$$

$$dy/dt = C \tag{4}$$

If $y = C/t$,

$$\int_{t_1}^{t_2} (C/t)\, dt = C \ln(t_2 - t_1) \tag{5}$$

$$dy/dt = -C/t^2 \tag{6}$$

If $y = Ce^{-gt}$,

$$\int_{t_1}^{t_2} Ce^{-gt}\, dt = (C/g)(e^{-gt_1} - e^{-gt_2}) \tag{7}$$

(If t_1 is zero, e^{-gt_1} is 1. If t_2 is infinity, e^{-gt_2} is zero.)

$$dy/dt = -Cge^{-gt} \tag{8}$$

If $y = Ce^{+gt}$,

$$\int_{t_1}^{t_2} Ce^{+gt}\, dt = (C/g)(e^{+gt_2} - e^{+gt_1}) \tag{9}$$

(If t_1 is zero, e^{+gt_1} is 1. If t_2 is infinity, e^{+gt_2} is infinity, as is the entire integral.)

$$dy/dt = Cge^{+gt} \tag{10}$$

If $y = Ct^N e^{-gt}$,

$$\int_0^\infty Ct^N e^{-gt}\, dt = C(N!/g^{(N+1)}) \tag{11}$$

$$dy/dt = CNt^{(N-1)}e^{-gt} - Cgt^N e^{-gt}$$
$$= Ce^{-gt}t^{(N-1)}(N - gt) \tag{12}$$

Note

For functions consisting of the sum of a series of terms the integrals or derivatives of each separate term are additive. For example,

$$\int_{t_1}^{t_2} (C_1 e^{-g_1 t} + C_2 e^{-g_2 t} + \cdots + C_N e^{-g_N t})\, dt$$

$$= \frac{C_1}{g_1}(e^{-g_1 t_1} - e^{-g_1 t_2}) + \frac{C_2}{g_2}(e^{-g_2 t_1} - e^{-g_2 t_2}) + \cdots + \frac{C_N}{g_N}(e^{-g_N t_1} - e^{-g_N t_2})$$

REFERENCES

Aseltine, J. A., "Transform Method in Linear System Analysis." McGraw-Hill, New York, 1958.

Baker, N., R. A. Shipley, R. E. Clark, G. E. Incefy, and S. M. Skinner, *Am. J. Physiol.* **200**, 863 (1961).

Bassingthwaighte, J. B., *Circ. Res.* **19**, 332 (1966).

Bassingthwaighte, J. B., *Science* **167**, 1347 (1970).

Bassingthwaighte, J. B., R. E. Sturm, and E. H. Wood, *Mayo Clinic Proc.* **45**, 563 (1970a).

Bassingthwaighte, J. B., T. J. Knopp, and D. U. Anderson, *Circ. Res.* **27**, 277 (1970b).

Bergman, E. N., D. J. Starr, and S. S. Reulein, Jr., *Am. J. Physiol.* **215**, 874 (1968).

Bergner, P. E., *Acta Radiol.* **Supp 210**, 1 (1962).

Bergner, P. E., *Theoret. Biol.* **6**, 137 (1964).

Berman, M., E. Shahn, and M. F. Weiss, *Biophys. J.* **2**, 275 (1962a).

Berman, M., M. F. Weiss, and E. Shahn *Biophys. J.* **2**, 289 (1962b).

Berman, M., E. Hoff, M. Barandes, D. V. Becker, M. Sonenberg, R. Benua, and D. A. Koutras, *J. Clin. Endocrinol.* **28**, 1 (1968).

Brownell, G. L., M. Berman, and J. S. Robertson, *Int. J. Appl. Rad. and Isotopes* **19**, 249 (1968).

Cuénod, M., and A. Durling, "A Discrete-Time Approach for System Analysis." Academic Press, New York, 1969.

Donato, L., C. Giuntini, M. L. Lewis, J. Durand, D. F. Rochester, R. M. Harvey, and A. Cournand, *Circulation* **26**, 174 (1962).

Gallagher, T. F., D. K. Fukushima, and L. Hellman, *J. Clin. Endocrinol.* **31**, 625 (1970).

Giuntini, C., M. L. Lewis, A. Sales Luis, and R. M. Harvey, *J. Clin. Invest.* **42**, 1589 (1963).

Gurpide, E., J. Mann, and E. Sandberg, *Biochem.* **3**, 1250 (1964).

Gurpide, E., and J. Mann, *Bull. Math. Biophys.* **27**, 389 (1965).

Hetenyi, G., Jr., and D. Mak. *Can. J. Physiol Pharmacol.* **48**, 732 (1970).

Høedt-Rasmussen, K., E. Sveinsdottir, and N. A. Lassen, *Circ. Res.* **18**, 237 (1966).

Holzbach, R. T., R. A. Shipley, R. E. Clark, and E. B. Chudzik, *J. Clin. Invest.* **43**, 1125 (1964).

Huff, R. L., D. D. Feller, O. J. Judd, and G. M. Bogardus, *Circ. Res.* **3**, 564 (1955).

Keating, F. R., Jr., M. H. Power, J. Berkson, and S. F. Haines, *J. Clin. Invest.* **26**, 1138 (1947).

Kety, S. S., *Pharmacol. Rev.* **3**, 1 (1951).

Kety, S. S., *Meth. Med. Res.* **8**, 223 (1960).

Kinsman, J. M., J. W. Moore, and W. F. Hamilton, *Am. J. Physiol.* **89**, 322 (1929).

Kowarski, A., R. G. Thompson, C. J. Migeon, and R. M. Blizzard, *J. Clin. Endocrinol.* **32**, 356 (1971).

Lassen, N. A., *J. Clin. Invest.* **43**, 1805 (1964).

Lassen, N. A., and A. Klee, *Circ. Res.*, **16**, 26 (1965).

Lewallen, C. G., M. Berman, and J. E. Rall, *J. Clin. Invest.* **38**, 66 (1959).

Meier, P., and K. L. Zierler, *J. App. Physiol.* **6**, 731 (1954).

Möller, E., J. F. McIntosh, and D. D. Van Slyke, *J. Clin. Invest.* **6**, 427 (1928).

Newcomer, W. S., *Am. J. Physiol.* **212**, 1391 (1967).

Orr, J. S., and F. C. Gillespie, *Science* **162**, 138 (1968).

Parrish, D., D. T. Hayden, W. Garrett, and R. L. Huff, *Circ. Res.* **7**, 746 (1959).

Pearlman, W. H., *Biochem. J.* **67**, 1 (1957).

Rittenberg, D., and R. Schoenheimer, *J. Biol. Chem.*, **121**, 235 (1937).

Riviere, R., D. Comar, C. Kellershohn, J. S. Orr, F. C. Gillespie, and J. M. A. Lenihan, *Lancet* **1**, 389 (1969).

Robertson, J. S., *Physiol Rev.* **37**, 133 (1957).

Sapirstein, LA., and E. Ogden, *Circ. Res.* **4**, 245 (1956).

Sheppard, C. W., " Basic Principles of the Tracer Method." Wiley, New York, 1962.

Shipley, R. A., R. E. Clark, D. Liebowitz, and J. S. Krohmer, *Circ. Res.* **1**, 428 (1953).

Shipley, R. A., E. B. Chudzik, A. P. Gibbons, K. Jongedyk, and D. O. Brummond, *Am. J. Physiol.* **213**, 1149 (1967).

Shipley, R. A., E. B. Chudzik, and A. P. Gibbons, *Am. J. Physiol.* **219**, 364 (1970).

Skinner, S. M., R. E. Clark, N. Baker, and R. A. Shipley, *Am. J. Physiol.* **196**, 238 (1959).

Steele, R., *Ann. N.Y. Acad. Sc.* **82**, 420 (1959).

Steele, R., *Fed. Proc.* **23**, 671 (1964).

Steele, R., J. S. Wall, R. C. deBodo, and N. Altszuler, *Am. J. Physiol.* **187**, 15 (1956).

Steele, R., N. Altszuler, J. S. Wall, A. Dunn, and R. C. de Bodo, *Am. J. Physiol.* **196**, 221 (1959).

Stephenson, J. L., *Bull. Math. Biophys.* **10**, 117 (1948).

Stetten, D., I. D. Welt, D. J. Ingle, and E. H. Morley, *J. Biol. Chem.* **192**, 817 (1951).

Stewart, G. N., *J. Physiol.* **22**, 159 (1897).

Tait, J. F., *J. Clin. Endocrinol.* **23**, 1285 (1963).

Turner, R. C., J. A. Grayburn, G. B. Newman, and J. D. N. Nabarro, *J. Clin. Endocrinol.* **33**, 279 (1971).

Wall, J. S., R. Steele, R. C. deBodo, and N. Altszuler, *Am. J. Physiol.* **189**, 43 (1957),

Waterhouse, C., and J. Keilson, *J. Clin. Invest.* **48**, 2359 (1969).

Weitzman, E. D., D. Fukushima, C. Nogeire, H. Roffwarg, T. F.Gallagher, and L. Hellman, *J. Clin. Endocrinol.* **33**, 14 (1971).

Wollman, S. H., and F. E. Reed, *Am. J. Physiol.* **202**, 182 (1962).

INDEX